A Profession *of*
Hope

Books by Jenna Butler

Aphelion
Seldom Seen Road
Wells

A Profession *of*
Hope

Farming on the Edge of
the Grizzly Trail

JENNA BUTLER

WOLSAK
& WYNN

Cover image: Natalia Bulatova
Cover and interior design: Marijke Friesen
Author photograph: C. W. Hill
Typeset in Caslon Book
Printed by Ball Media, Brantford, Canada

The publisher gratefully acknowledges the support of the Canada Council for the Arts, the Ontario Arts Council and the Canada Book Fund.

Wolsak and Wynn Publishers Ltd.
280 James Street North
Hamilton, ON
Canada L8R 2L3

Library and Archives Canada Cataloguing in Publication

Butler, Jenna, 1980–, author
A profession of hope: farming on the edge of the Grizzly Trail / Jenna Butler.

Includes bibliographical references.
ISBN 978-1-928088-08-0 (paperback)

1. Butler, Jenna, 1980–. 2. Farm life–Alberta, Northern. 3. Organic farming–Alberta, Northern. 4. Frontier and pioneer life–Alberta, Northern. I. Title.

S522.C3B88 2015   630.97123'1   C2015-905099-5

For Thomas:
my partner in all things,
with awe and love.

For the land:
home, heart.

Farming is a profession of hope.
– Brian Brett

There are no unsacred places;
there are only sacred places
and desecrated places.
– Wendell Berry

Table of Contents

PREAMBLE

IN AN ERA OF LARGE-SCALE agribusiness and multinational corporations dedicated to the privatization and control of the world's seed stocks for profit, starting out as a small-scale farmer seems like an uphill slog, a thankless task. But there's beauty in it, found in the steadily growing community of supporters who are searching for more than just local food. They're looking for a way back to the land itself, its histories and pleasures, in a form far different than the gigantic feedlots, pumpjacks and canola fields that checker the North American prairies. They want a hands-on experience when it comes to food and to the land that produces it. Instead of needing a haz-mat suit to grub chemically drenched Yukon Golds out of the field, they want to be able to scrabble out their produce with bare hands. Even better, with the kids in tow.

In spite of the burgeoning support base for small farms in North America, there's still the wider issue of food security in a country such as Canada, where most of the people (and thus, the small farms) are located in a relatively limited southern belt. With overseas food shortages and domestic food prices on an ever more drastic rise as an environment under threat becomes increasingly unpredictable, there's an immediate concern regarding food production in Canada's colder climates.

This is where the northern small farms come in, in all their quirky glory. They're places capable of producing a great deal of high-quality food in a manner that doesn't take more from the land than the soil can bear. Small-scale organic farming methods offer ways of reclaiming marginal land whose goodness has been used up, putting into play farming methods that allow for a strengthening, rather than a wholesale destruction, of ecosystems. These are the spaces that protect and preserve the heirloom seeds passed down for generations and adapted specifically to the strictures of our local weather. At the same time, they're more than just cold-zone seed vaults: Canadian small farms are places in which the rough beauty of the landscape can be seen thrumming just beneath the surface of daily life. For the many who live on them, they represent a tying together of ideology, economics and functionality. These farms provide fiscally viable returns to the natural world in a diversified way that is no longer possible on many large-scale operations. Small farms are celebrations of what it means to live closely and responsibly with the land through all seasons, no matter what the weather or the economy throws your way.

Right now, the weather is throwing an eight-month winter our way. We found our farm during a winter much like this one back in 2006: the land was wild, overgrown and sunk in snowdrifts to the knees. We were captivated. We spent the dark months of 2006 snowshoeing the land before we signed to it, but we knew what we'd find come the melt in April. The quarter section (one hundred sixty acres here in Canada, rather cryptically listed on the deed as "160 acres, more or less") consists of one hundred thirty-five acres of northern mixed forest and boreal muskeg spruce (dwarf black spruce growing over deep peat bogs thick with Labrador tea). At the far corner stands a twenty-five-acre hayfield that one of the neighbours keeps in alfalfa for his dairy cows. When we found the land, we had no illusions. We knew that the only reason the quarter was going for a price that two teachers, living in a tiny rented apartment in the city, could afford was that it was rough northern bush. All of the big industrial farmers in the neighbourhood had looked at that land and turned it down. "Too much effort," they later told us. "Too much work to clear, and too much peat."

But where they saw wasted effort and backhoes sinking into acidic bog water, as would-be small farmers, we saw something else. The land had scarcely been touched, and while the access was poor and the ground springy with peat, the forest offered a built-in buffer zone from the conventional farms around us. The hayfield was beyond our reach at that point; at the far end of a rough old seismic line that turned into a swampy track in the spring, it was an ideal building site with dirt-poor access. We rented it out another year to the neighbour, who was happy to have the extra twenty-five acres after a bad year for hay the summer before. Every extra bale he could bring in was one fewer he'd have to buy. We turned his first rent cheque over to equipment and began nibbling away at the corner of the property that fronts onto the township road. A dead willow thicket, it promised easiest access for clearing. For gardening, it boasted earth that was deep and rich from years upon years of untouched willow-leaf mulch and decaying swamp grass.

A bucolic farm life? Nothing was farther from the truth that first year. The land was an hour and a half from Edmonton, and without a place to stay, we were driving out and back every Saturday and Sunday, pacing clearing work against full-time teaching jobs. The same dark, moist earth that promised a spectacular garden in future years yielded us an unwelcome bumper crop that first spring: a hatch of mosquitoes so bad that they literally coated us from head to foot the moment we emerged from the car. This was followed by a batch of nettles that reached five feet in the unrestricted root run of peaty soil.

This is not the story of a ready-made farm, complete with generations of history, carefully tended tools and sturdy clapboard farmhouse. Those came later, as we learned the stories of our county and added a cabin of our own to the land. That first summer, though, there was nothing to move into, and so we moved *out*. After our teaching jobs wrapped up for the summer, we spent every available moment out on the land. At its most basic, it was literally just us, an axe, a chainsaw and a quarter section of northern bush picked out in inquisitive moose.

This is what the small farm movement is all about. Inspiration. Diversity. Maybe a touch of madness: the desire to be out under the sky in all weather, to be working with our hands as much as possible, turning to big machinery only when necessary. When it comes down to it, it's about hard work and long hours put in with the knowledge that in order to found our farm from nothing, it's been necessary to hold other jobs in the wings, full-time jobs that also require energy. My husband and I are not extraordinary people. We're everyday folk, and we've lived in a big city for most of our lives. But there's a very specific love that drives us out here, that makes us want, more than anything, to be able to enrich our lives and those of others by working with this land, taking just what we need from a small corner while safeguarding the rest as a wild place for future generations. There's an excitement about it all, not just about small-scale northern farming, but about what it can mean *now*, at this point in time, for this country. Paul Hawken, entrepreneur and environmental activist, frames it exactly: "When asked if I am pessimistic or optimistic about the future, my answer is always the same: if you look at the science about what is happening on earth and aren't pessimistic, you don't understand the data. But if you meet the people who are working to restore this earth and the lives of the poor, and you aren't optimistic, you haven't got a pulse."[1]

Welcome to the farm.

CHAPTER ONE

Flipping the Switch: From the City to the Farm

WHAT MAKES US STEP AWAY from a stereotypical urban life? For some, perhaps it's a small departure: a handful of chickens in the backyard, a few raised beds for beans and greens or a rooftop garden if the condo building permits. For others, it's the whole hog: departing city life altogether for a different way of being. Either way, there's a shift that happens within some of us, and the desire for change – to be closer to the earth, our food and the seasons – becomes a necessity.

There's a moment when, no matter who you are, if you've been "dreaming out," something flips that switch. You make the transition from a life of *what if* and *wouldn't it be great if* to *why not?* And more importantly, *why not now?* No matter when that happens, it's a date that stays with you: the moment you granted yourself permission to have that life; the moment that everything changed.

Our moment came during the early winter of 2006. After months of driving isolated back roads outside Edmonton, my husband, Thomas, and I drew up at a single-strand wire gate on an early November evening. It was five o'clock and dark as the inside of a barrel. The day's exploring had taken us longer than we'd anticipated, and we'd gotten to the circled point on our map long after the sun, wandering toward the solstice, had set. We shivered our way out of the car and stood in the glow of the headlights, looking up at the weathered grain bin that marked the entrance to the property. Out of nowhere, a pack of coyotes began its evening chant. A great horned owl called in the deep woods, and a second one answered. Thomas squinted up at the metal bin, reading the paint. "Look at that," he laughed. "It's got your name on it." It was one of those old Butler bins. We paused and considered each other.

Like that, the switch flipped.

Why were we even out there at all, two lone figures on a snow-fringed township road in early winter? Like many people, we'd dreamed of having a little plot of our own, a small piece of land or a double lot in the city where we could grow our own food, but it had been a semi-formed dream. Then we moved to England for a year – my husband on sabbatical from teaching, me working on my Master's in poetry – and for ten months, we settled in the northeastern part of the country where my father's family hails from. There, surrounded by the small rural villages, intergenerational farms and large country gardens of my early childhood, I rediscovered that land with my husband, and we became aware of a desire to have something like that for ourselves. But, being stubborn and steadfast Canadians, we wanted to try to build a rural life back at home in Alberta, not in the tiny village where my grandmother lived. Beyond just the homing instinct, there was something about small-scale farming in a cold climate with a short growing season that appealed to us. The challenge beckoned.

Michael Pollan speaks of a formative moment "a few years shy of [his] fortieth birthday [. . .] when the notion of a room of [his] own, and specifically, of a little wood-frame hut in the woods behind [his] house, began to occupy [his] imaginings with a mounting insistence."[1] Fiction writer and small farmer Barbara Kingsolver lists a number of practical reasons for moving her family away from an urban lifestyle, including being closer to extended family and loving the wild life spent outdoors. Mostly, though, she came to realize that her family's move emerged from a desire to be nourished by the ground they lived on: "[They] wanted to live in a place that could feed [them]: where rain [fell], crops [grew], and drinking water [bubbled] right up out of the ground."[2]

For Thomas and me, both city-based teachers, both as prepared as we could be for an adventure of such magnitude, the thought of finding a piece of land to protect from development was deeply attractive. In 2005, when we came home from England, where any construction is years in the planning before spade is set to ground, we found that much of the best black-soil land immediately around Edmonton was being sold off for development. Industrial parks and suburbs were being built overtop of some of the province's most fertile farming belts. Knowing we couldn't afford even half an acre in those developments (and who wants to – or can – shell out a cool million for a lot next to an industrial park?), yet still wanting to purchase a wild space we could keep safe, we searched farther and farther afield. For a year and a half, our weekends were composed of very pointed road trips: the two of us in our small black Jetta, equipped with mugs of tea and a well-circled map, buzzing around the back roads amidst farm trucks and mobile drilling rigs and snowplows. We learned that the farther we got from the city, the lower the price of the land, but with those cost benefits came a host of new problems: groundwater so contaminated from nearby oil and gas wells that farmers could light their methane-filled tap water on fire, and muskeggy dirt that could suck a one-ton truck down during the spring thaw and not release it until freeze-up. Everywhere we turned, there was gas flaring and quad damage and quarters that had been logged within an inch of their lives.

Over the years, many people have asked us whether we came from farming backgrounds, and if that was how we knew to look for the right sort of land. There might be some farming blood in me – my father's side of the family has lived and worked in England's farming country for generations – but Thomas is a city boy, born and bred. To be honest, my biggest experience of farming as a kid was going to the annual Harvest Fair down at Fort Edmonton Park after we moved to Alberta. So when people ask me whether we knew something beforehand about farming, as though there's some secret skill set that allows certain folk to make the jump from the city to rural life with a greater deal of success, my answer is no. That's exciting and terrifying, isn't it? That means there's no magic pill for success, no easy way to go about it. Anybody with determination and knowledge can do a pretty decent job of getting back to the land.

Our decision to buy the piece of land that would become our small farm was one born in the bones. For me, it echoed with memories of the tiny farming community I come from in northeastern England, where some of my family still grows sugar beet and grain, and raises cattle. Although I don't recall much of the life I left behind in England, I've always harboured a deep-seated desire to find a piece of land outside the city to call home. Thomas, too, born in the Netherlands and raised in Edmonton, has always loved being outdoors; he's a devoted backcountry hiker and camper, and has hauled me up a fair number of mountains. We both love being *out*, whatever the weather. I can't help but think you're asking for trouble if you make a break from city life when you don't really like being in the country.

This means being able to stand up to *everything* the country throws at you: in our case, windstorms, lightning strikes, mosquitoes, flood years, drought years, raiding moose, slumpy peat soil and stinging nettles. I have dear friends who can't stand cramped city life but whose only appreciation for the country comes from running their snowmobiles through it in the winter. If that's you, all I can say is look for a double lot in the city and get your taste of rural living from having a big, beautiful garden, because jumping ship to a country property isn't going to be the answer if you only love the land at specific times of year. You have to want to be out, and then for as much of the day as there is light, especially if you're running a farm. When friends learned that we were living in a fourteen-by-six-foot truck camper for four months of the year during the first few years of the farm, they were horrified. "How can you *survive* in there?" they'd ask, looking at their own partners. "We'd kill each other!" The answer was, we were outdoors. Thomas built a huge gravel patio outside the camper, and this was our living room for those four months. Every day, we were out of the camper at dawn and in again only to sleep, or to escape the rain and have a cup of tea. The world outside, the world of farm and forest, became our everything.

What began as a practicality – we simply didn't have the money to buy close to the city – changed along the way into something more profound. We came to the idea that what we really wanted was to find a tract of land far outside the city that we could protect against future urban development, and that we could work with to ensure our own health and survival. We wanted a flower garden, a big one, and an orchard with beehives. A cabin, eventually, as a home. A large hayfield for growing crops, and maybe one day for growing the bales for a straw bale house. A spread of forest to manage sustainably as a woodlot. We wanted the good health that came from working hard outdoors in all weather. And we began to understand, as I suffered from progressively worsening allergies when eating conventionally farmed, store-bought vegetables in the city, that we needed to completely change the way we ate.

So, that November dusk in 2006, when we stood by the side of a small township road in northern Alberta's Barrhead County, an hour and a half's drive northwest of Edmonton, and listened to the owls and the coyotes, we were aware of a very real separation beginning to happen between ourselves and the city. The switch had been flipped. The land we stood on was relatively untouched, distant enough from oil and gas, and full of animal life. We didn't know what the ground would look like come summertime, and we realized that we stood at the base of a huge learning curve, but we were already sold. Somehow, over the next few years, we would find a way to make that land our home.

And truly, once doubting and worrying and hedging your bets are set aside, just committing to the decision to live closer to the land is most of the battle won.

CHAPTER TWO

Bug Off

REMEMBER WHEN I SAID that if you're planning to dream yourself a country life, you'd better be able to deal with everything the wilderness throws at you? It's the honest truth. And much of the time, what gets thrown at you is bugs.

We call our place Larch Grove Farm for the tamarack trees that fill its forest, but that first summer back in 2007, there was nothing recognizably farm-ish about it. Part of what attracted us to the land was its nearly untouched nature: it had never been clear-cut or used for cropping. The elderly couple who sold it to us as the final quarter section of a once-thriving family farm had used it as a hunting ground and midden for years, so we inherited a worn grain bin full of five-foot spans of moose antlers and a one-hundred-thirty-five-acre forest picked out in rusted snowmobiles, decaying upright pianos and ruined wagon wheels. And *bugs*.

We'd fallen in love with the property during the deep winter, but thank goodness we'd had sense enough despite our infatuation to get out the snowshoes and walk the land before entering into the purchase agreement. We learned pretty quickly that the same land that was home to great horned owls and a den of coyotes was also home to one hundred thirty-five acres of muskeg spruce forest and Labrador tea – in other words, not your run-of-the-mill farmland. The quarter we'd fallen for lies at the southern edge of the boreal mixed-wood region, a zone of white spruce, poplar, black spruce and muskeg swamps that runs right up to the far northern border of the province. Our farm is several miles off the Grizzly Trail, formerly the original Klondike Trail, a secondary highway that bisects the town of Barrhead and runs straight up into the North. It's beautiful country, a combination of rolling hills and deep peat bogs. Our quarter in partic- ular (as we later found out from the hydrogeologist we brought in to conduct a groundwater survey) is home to both a prehistoric lakebed and a riverbed, and is at the bottom of a bowl of hills. All of that adds up to two things: first, because of the variable terrain along the Griz- zly Trail, we're gardening in a frost hollow, one of the places most rural folk who know better run screaming from. Second, the peaty soil stays damp year-round. And that means – you guessed it – bugs.

THAT FIRST SUMMER, we pulled up by the grain bin after our ninety- minute drive northwest from Edmonton, intending to start a bit of clearing in the dead willow thicket near the run-down tool shed. Moments after we climbed out of the car, we were met with a cloud of mosquitoes that defied all our city-bred senses. Like everyone else in prairie cities across western Canada, we'd taken for granted the hordes of park workers with backpack sprayers and the low-flying fogging aircraft that took care of the mosquitoes in the public parks and river valleys. We didn't like the procedure, and I was always prone to bad allergies after the city had sprayed, but we didn't doubt that it worked. We'd never encountered mosquito swarms of this magnitude before, and they were *hungry*. Even after we piled back into the car some twenty awful minutes later, they landed on every closed window, buzzing after blood. It was Hitchcockian.

In true hobby farmer fashion, we decided to abandon ship and head out on a road trip to Vancouver Island instead – miraculously mosquito free. We came reluctantly back to the farm the next month to see if the bug problem had resolved itself. Astonishingly, it had. The peaty soil that turned the ground into a quagmire of marsh marigolds every spring had dried sufficiently in July's heat to become something resembling ordinary garden soil, though a little on the dusty side. The willow thicket was a mess of broken grey limbs and waist-high marsh grass, but compared to the previous month, it was relatively bug free.

We learned an invaluable lesson in those first two months at the farm: if we wanted to be able to work on the land and not be eaten alive, we were going to have to do some serious clearing. Not only was the willow thicket a major fire risk right there at the entrance to the property, but the boggy ground beneath it would be a perennial problem. As long as the soil was spongy and wet, we'd be greeted at the start of every summer with a bumper crop of mosquitoes.

And so the Summer of Mosquitoes led into the Summer of the Chainsaw. Well, not immediately. In true idealistic, back-to-the-lander form, we who had been living amongst the pages of *The Good Life* and *The Harrowsmith Country Life Reader* decided to start clearing the willow thicket with our lone tool at the time: an axe. That lasted for a day. The first evening, after having spent hours ricocheting his axe off a springy, eight-trunked Bebb willow, Thomas presented me with his double-handed set of blisters and announced, "We're getting a chainsaw."

From an axe to a chainsaw, and from there to a secondhand farm truck and pull chains for stumping the next summer, we slowly nibbled away at the one edge of the property accessible from the township road. As the dead willow thicket came down, the ground dried and the lay of the land emerged, and we realized that the hordes of mosquitoes were a thing of the past. Not because we had changed the land enough to be inhospitable to them, but because more songbirds sought out the newly opened space of the farm garden to feed. Where once we had a handful of chickadees and woodpeckers hammering away on the dead willow, we now also had white-crowned sparrows, warblers and goldfinches, whiskey jacks and ravens. A broad-winged hawk moved in to raise her brood near this surprising new twenty-four-hour food mart, and she returned to the woods every summer afterward. The owls thrived, and dragonflies appeared seemingly out of the very air to help the songbirds keep the mosquito count down.

Farming, though, is a constant font of the unexpected. It was early June, university classes were over, and I'd already been putting in long hours in the farm's market garden, getting the early seedlings transplanted from cold frames to garden beds in advance of the growing season. Thomas was a mere four weeks off the end of the school year and actually looking forward to starting his two-month stint as an unpaid farm labourer, weeding said beds under my watchful eye. We were humming down the township road that dead-ends at our gate when something zinged against the side of the Jetta. Something else followed suit. Thomas slowed down, and in that instant, listening to the chitinous pings off window glass, we realized: *grasshoppers.* The front field, leased out by our neighbour to the farmers down the road, had been sown in wheat that year, and when we turned to the field, we saw that the just-forming stalks were alight with hoppers. It was an infestation.

As the summer progressed, the wheat field and the township road were so thickly coated with grasshoppers that we couldn't walk a step without causing a sort of ground-born hailstorm. So that turned out to be the Year of the Hoppers, and by the end of the growing season, we and all the farmers around us realized that Larch Grove Farm had fallen under the fickle spell of sheer dumb luck. We were out in the garden one hot August afternoon when the contingent of farmers who leased the front field showed up in their truck to check on how the wheat was heading up. There were the usual few minutes of rumbled talk in our driveway, but this time the men at the gate stood around and toed the gravel, shaking their heads and looking helplessly down at their hands. We were unfolding ourselves from weeding position to go and say hello when one of them saw us in the garden and waved, and they all came on over.

These men have farmed for generations, and they've turned the front field for many, many years. They know better than just about anyone that our quarter is wasteland, an untouched forest because of the muskeg that lies beneath it, land that no local in his right mind would pick up for farming. It'd be madness to clear it, and add to that its frost hollow location ... it stacks up as some kind of undesirable. Knowing this, imagine their faces when they walked into our small market-garden compound and found that the grasshoppers had barely touched the rows of food crops. They told us in short order that the hoppers had ruined the wheat crop and taken out all the farm gardens in the area, open as they were to the fields. We, however, had kept a thick shelterbelt of willow and poplar between ourselves and the surrounding fields, mostly because we wanted to avoid chemical drift when the farmers sprayed their crops.

What we *didn't* know – what they soon set us straight about – is that grasshoppers attack a field by walking in. When there's a shelterbelt in their way, it takes them too long to make their way through it, so they tend to go for easier pickings. By the time the summer had worn on and a few grasshoppers finally appeared in our garden, reaching the airborne stage of their lifespan and flying in, they didn't do much damage. An adult in that stage of its life has lost much of its appetite. With that simple conversation, we started to understand just how small-scale farming and organic methods were helping us to keep our land (and our garden) safe, even if unexpectedly. If we hadn't had that shelterbelt to protect our vegetable garden against the chemical drift from next door, the hoppers would have cleaned us out like everybody else.

The visiting farmers marveled at the size of our beets and carrots. They wandered around the rustic garden we'd carved out of the frost hollow, perplexed at the newbies and their strange luck. Finally, one of them turned to Thomas and said, shaking his head, "I don't understand it! There's hardly any damage here at all. I know you've got a good shelterbelt, but geez. We tried *everything* on our garden to kill the grasshoppers, but they ate the whole lot. What did you use?"

Thomas thought briefly and then replied, delivering with absolute sincerity the answer that has branded us the weird teacher hippies ever since: "Worm poop."

CHAPTER THREE

Black Gold

OUT IN RURAL NORTHEASTERN England where I was born, a problem in the crop always came with a host of options and a variety of colourful names: Paris green bait for grasshoppers and rats; paraquat and diquat to burn off the tops and harden the skins of potatoes for mechanical harvesting; a slew of neonicotinoid pesticides for the rapeseed crops. Here on the Canadian prairies, I've watched the same handful of die-hard bug-busters at work over the years, and I've seen the bees come in disoriented from a visit to the canola field, scrubbing at the guard hairs across their bodies as the chemical sprays adhere and absorb and disrupt their ability to fly.

I'll stop here and interject that there are always going to be those who will fight alongside Big Ag and say that neonics aren't the silent killers they're accused of being. I want to clearly acknowledge that we don't yet have all the facts – and here I'm talking about the decades-long studies – to prove the ultimate effects of neonics on our bee populations, just as we don't yet have the long-term studies to prove conclusively that genetically modified (GM) crops are without human health repercussions. What I can turn to is the evidence from my own garden, and what I've observed is that the bees coming in off the chemically drenched fields around spraying days are in bad shape, disoriented and obviously suffering some discomfort. I suppose it'd be the equivalent of running blithely through a glyphosate sprinkler midsummer. As a gardener of some twenty-five years, when the studies I need aren't in, I turn to what I can find: the words of other farmers, the historical records and the observations of my own crops. I'm fairly certain that a concentrated chemical drench isn't good for any of us, and when you add to that the vastly changeable winters of recent years and bees stressed from varroa mites and long-distance travel to pollinate thousands of acres of flowering crops, you have the perfect recipe for a full-system failure.

Our loyalty to organic gardening on the farm has presented us with an incredible learning curve. We'd been used to instant solutions in the city: when planting annuals, reach for a bucket of quick-release fertilizer; when fighting black spot on roses, reach for a nefarious pink canister of leaf dust; when the slugs get to be too much in the broccoli, hunt down the utility-sized plastic jug of slug pellets. (Also, keep your cat inside for the next week, and remind the neighbour that her kids aren't safe in the garden patch until the rain's watered the pellets in.)

But over the years, and especially in the five that led up to the purchase of our farm, we began to understand just how the buildup of chemicals in the body eats at you. I am the one in our family with the chemical sensitivities, and it became second nature to me to eat at a restaurant and emerge hacking and wheezing because of a tainted salad. I came to unhappy terms with the fact that anything including iceberg lettuce, often treated for salmonella in commercial crops, caused me to welt up in hives. By the time we found the land in 2006, I hadn't eaten an apple in close to fifteen years because of the painful coughing that ensued after consuming the waxes and pesticides caught up in the apples' skins. Increasingly, I found myself sick and frustrated with the very food that was supposed to support good health. I desperately needed to make a change.

It's a frightening thing to leave behind the chemical-filled world of "perfect" food that we grow up with in the West and go the organic route, where you are very much aligned with nature. We are, on this continent in particular, great fans of the quick and easy. And so maybe you come to organics because of a fascination with the arguments behind it, or maybe you come to it because you realize that your sleek child-body has grown up to become slow and sluggish and puffy with residues, but somewhere along the line, you start to understand that there's a certain sense to food-growing that takes only natural elements out of the soil.

Our poisoned food is deeply symbolic of our skewed way of interacting with the land around us. I find myself thinking of Wendell Berry, the American poet-farmer, and his insistence that "food, rather than simply being fuel, is the most concrete and intimate connection between ourselves and the earth that exists."[1] Rather than getting up on a soapbox, my way of managing this discussion has always been to bring in a box of sweet summer carrots, just picked in the market garden and rubbed free of soil. Food like that, in which you taste the sugars and the sun, and which, if you're like me, you bite into gingerly in anticipation of an allergic reaction but then come away from unhurt, is its own kind of revelation.

I READ BERRY long before the practicality of his words made sense to me as a farmer, the admonition that, "in many cases, human progress has been measured in terms of our ability to combat and control the land, and so to establish human flourishing in opposition to the well-being of the earth."[2] This opposition takes a variety of forms: the clearing of vast fields for monocropping, the application of herbicides shortly after seeding and pesticides while the crop grows, and the development of agricultural corporations that care nothing for biodiversity, small farms or organic methods. Before we'd sown our very first crop at Larch Grove, a handful of seed potatoes, Thomas and I found ourselves worrying over the question of how to replenish the soil without adding things to it that would harm us, and the land we'd bought to protect, in the long run.

Back in 2008 while we were farming, Thomas bought a bucketful of red wigglers from our local Earth's General Store with the intent of teaching his Grade Four students about worms and how they work in the soil. Suddenly, we were the caretakers of a thousand or so crawlies living in a pail under our kitchen sink, and we came to understand within the first week just how great those worms were at composting. Every scrap of vegetable material we had, they ate. Spoiled sprouts. Carrot peelings. The tops and tails of string beans. We watched in something between astonishment and alarm as they chewed their way through whatever we could throw at them.

We'd noticed a lack of worms in the soil at our farm since the first summer we broke ground. The land was all peat, and the dusty summer garden beds barely held up the carrots and beets, the soil threatening to pick up and blow away in the first good wind. Northern peat bogs are so densely packed with roots and fibres under the surface that there's no room for worms or mineral soil, and when hoed, the land turns into a dust crust that breaks up and blows away as soon as it dries. In the winter, the ground freezes into something resembling firebrick. We soon learned that the only thing turning the soil in that garden was us.

At school, the worms were installed in a gigantic Rubbermaid tub in the storage locker and fed daily on the leftovers from the school's fruit program. In a few short weeks, we began to see our first unexpected windfall: a harvest of worm castings that the books told us would enrich our soil faster, and in a more nutrient-balanced way, than anything outside of compost. Everything in the market garden was planted with a handful of worm castings on the side. I made gigantic "tea bags" of castings to steep in pails of water and use as foliar feed. That earthy, dark water smelled of deep underground, and it spurred plants to growth as though they'd been shot up with steroids overnight. Suddenly, and with very little effort, we had obtained a way of fertilizing the food and flower crops at the farm that didn't involve questionable pesticides or caustic ingredients. In later years, as we went fully organic, so did the worm bins.

Worm castings, or "worm poop," as Thomas's kids refer to the black gold, are easily produced but not meant to be shovelled wholesale into the garden. They kill crops outright when we do that, as we discovered in the first years: it's a Henry VIII sort of thing, the plants overdosing on rich food and smothering in the fat, greasy soil. But incorporated into planting holes and pots, scattered around plants or watered in as tea, worm castings do something that no chemical fertilizer can. Within a day of their application, colour returns to drooping plants. Tomatoes and peppers settle down to fruit in the greenhouse. Disease in plants isn't magically removed, but it's reduced to the point at which the plant's own system can take over.

We still raise the worms from that first year's batch, though they've been divided and divided over the years, moved on to larger and more numerous tubs, with colonies split off for friends who want to try vermicomposting at home with their kids. In just a handful of years, we've watched the soil in our market garden completely turn around by the addition of worms. It no longer reduces to powder in the summer; our garden is no more a miniature Dust Bowl after July windstorms come through. The soil tilth has improved and opened, and in the early spring, the robins take over the tilled, planted beds to haul fat worms from the ground. Red wigglers don't often survive the winter in our cold Zone 2 garden, but yearly use of worm compost sees them thrive, in addition to earthworms, throughout the garden in the warmer months.

Yes, the neighbouring large-scale farmers think we have a screw loose. And yes, some friends still wince when they learn that we have a worm composter under the sink or in the basement in the city. But the benefits to the soil, the health of our plants and, ultimately, our own well-being are beyond price. We haven't reached for slug pellets or a bucket of fertilizer for well over a decade now, and I can safely grab for anything growing in the garden while I'm out weeding and graze away without fear of allergies. Our friends' children never tire of playing in the worm bins, and Thomas still teaches soil science with a fresh batch of red wigglers every year. Both through teaching and through our own practice, we're interested in sharing a very simple solution to the age-old problem of how to fertilize the plants that produce the majority of our food in the year. And when you come down to it, there's something inherently comforting about reaching for a solution that doesn't come with a cryptic back panel of poisonous ingredients and cautions.

Berry speaks, in *The Unsettling of America*, of our willingness to destroy the earth as a reflection of a parallel willingness to injure ourselves: "To damage the earth is to damage [our] children. To despise the ground is to despise its fruit; to despise the fruit is to despise its eaters. The wholeness of health is broken by despite."[3]

No more chemicals on this land. We got rid of the last wilted sack of insecticide left over from the old farmer who owned the property before us when we cleaned out the junk beside the tool shed. And, watching the heirloom tomatoes rise into seven-foot giants in the greenhouse after a good drink of worm-casting tea, we'll never look back.

CHAPTER FOUR

Everything is Coming Up ... Weeds

WHEN I REREAD BRIAN BRETT'S astonishing *Trauma Farm* or come back to Thoreau's *Walden*, I wonder how they dealt with that scourge of the summer garden: weeds. Thoreau, all seemed to be coming up roses in your bucolic bean field; I sit reading your words by the campfire and wonder how you managed pigweed, stubborn wild mustard, five-foot nettles, marsh grass. Or those years when the garden floods with spring snowmelt or May rains, and the bulrush seeds come in from God knows where. Within days, we're inundated.

This is where local knowledge runs up hard against our city brains and stubbornness. Nobody gardens in a frost hollow; likewise, nobody in their right mind starts a market plot in the midst of one hundred thirty-five acres of wild forest. The untouched land will always be troubling the edges, pressing our bounds.

The first year we began tilling the soil, we realized, as a thin green haze covered the potato beds (and later, as our hands became covered with angry red welts during weeding), that our tilling had freed up several generations of nettle seeds. We fought Canada thistle outbreaks so bad that we honestly considered renting a goat from the neighbouring farm to snack away the problem. In dry years, we combatted pigweed and wild mustard; in wet years, chickweed and horsetail. We learned by trial and error (oh, the error!) that if you run a tiller over a bed spattered with horsetail and lungwort, you will soon have a bed of *nothing but* horsetail and lungwort. We learned that chickweed can mature its seeds on a hot compost pile weeks after you've pulled it from the garden, and that poor man's orchid can sprout its seeds underwater, spontaneously. The old adage is that only cockroaches and lawyers will outlast the end of the world, but I'd have to add chickweed to the list.

The old farmer we bought the property from used to joke that nettles were the best thing in the world to help him with his swelling joints. "You just gather up a nice bunch of nettle tops and flog yourself with 'em," he'd joke with a wink. "Pain's so damn bad that you forget about anything else. Best arthritis treatment I know." Curiously enough, he's right: the mild skin irritation and increased blood circulation from the nettles actually overrides the greater pain and irritation of the arthritis (but at the time, we just thought he was quaint and kind of nuts). Nettle flails aside, we discovered that first year that if we till the soil and then fall behind on the weeding, the nettles will be taller than we are by the start of August.

And so what else is there to do but to turn the weeds into things that benefit the garden? We don't have the option of Roundup® or 2,4-D (and no part of Agent Orange was really an option for us anyway), so all the weeds that rear their derisive heads in the summer garden go to either food or fertilizer. I've learned that nettles, picked young, make a delicious cream soup; older, they go into the batter for squash-blossom fritters or are dried en masse for tea. Before the gigantic plants go to seed, I whack them down with a billhook and lay them on the compost pile.

The garden writers I love best – Lois Hole and Wendell Berry, Barbara Kingsolver, Patrick Lima, Eliot Coleman – are all very aware that the garden is a cyclical creature in its appetites, and that anything it sprouts can, with some care, be turned into food to nurture it. Weedy grass clippings are laid out on tarps in the sun to dry into mulch or brown layers for the compost pile; wild mustard is harvested for its zesty greens. I haven't yet found any use for horsetail, though, and the chickweed still threatens to take over the strawberry and asparagus beds every August. It makes a good medicinal tea for anyone suffering from irritable bowel syndrome, but nobody I know has gut trouble on the same scale as we have chickweed! Wild raspberry sends its madcap runners through everything – easily controlled with the brushcutter after July's harvest of tart, tiny berries.

When I read Wendell Berry, I'm struck both by the lyricism of his words and by their tendency, in the present day, to be attributed to a bucolic way of life that ended with the Great Depression, or perhaps earlier, with industrialization. He speaks of how he came to perceive his body and his routines "as brief coherences and articulations of the energy of the place, which would fall back into it like the leaves in the autumn."[1] In this, I am reminded that no matter how severe the tide of interlopers the garden faces, it's not my place out here to contaminate this land with anything artificial. We're fortunate enough to be living on this land, working with it to grow our food and craft a low-needs way of life, but we're not permanent fixtures here. We'll have our handful of decades, and then we'll be gone. It's not our place to be putting anything into the soil that's going to have a long-term negative impact on the way this land can sustain itself, and that's where we are very different from conventional farmers. The organic approach doesn't look at the land in terms of the crops that can be gotten from it this year or next, or even five years down the road. Our approach looks at leaving this land in a better state than we found it for the people who will come after us – we think in terms of generations. It's a difficult binary, the desire to care for the land and, at the same time, the knowledge that that particular care requires an absence of some of the materials that might make tending easier in the moment. I stare down the thickets of nettles nodding out of flower, dropping their little seedlets over the water barrel and sneakily attempting an infestation every time I dip a can, and I understand Berry's confession that "here, in the place I love more than any other and where I have chosen among all others to live my life, I am more painfully divided within myself than I could be in any other place."[2] There is no "easy out" in this sort of caring, as there is no easy out in any deep sort of love.

And so I do what I can. Spruce seedlings are potted up and moved into the shelterbelt. Little birches are shifted back into the woodlot where they'll provide firewood in fifteen years instead of cluttering up the onion beds. Twice a summer, once when the willow catkins fuzz out and again when the poplar blows, I grit my teeth and pull on my gardening gloves to release the vegetable beds from a sudden carpet of taprooted seedlings. Everything cycles through the compost pile except for the truly noxious weeds and any plants suffering from disease. Everything is returned to the garden.

The compost pile itself is rather a thing of beauty. (I think of New England gardener and painter Tasha Tudor at this juncture: her veneration of the everyday, the sustaining, and in particular, a lovely compost heap). When we closed the loop on garden care and decided that even bringing in cow manure from the dairy farm a few miles down the road was an unacceptable risk (too many hormones, too much unknown), the garden had to become its own best food source. And so the compost pile grows like some bizarre structure outside the moose fence, towering to almost six feet in the height of summer and thoroughly blasting itself in the heat of July. By the time the summer has run its course, the compost, watered weekly in hot weather and turned, has broken down into something resembling a gigantic chocolate cake, earthen-smelling and crumbly.

Everything goes into it. Nettles, before seed. Grass clippings, dried to add mass. Small twigs, cabbage tops (which the deer will insist on pulling out and eating in the dead of winter), the contents of deadheading bins. Peelings, wood ashes from the stove, bio-char from the burn bin to add carbon back into the soil. Autumn leaves, shredded with the lawnmower to decay down over the winter into leaf mould. We're obscenely proud of the compost pile. It's an emblem of how nothing is wasted here, how everything, weed and otherwise, can be cycled back into the earth.

That compost, cooked in the sun's heat until all suspect seeds have gone inert (except chickweed, alas), is the stuff that bulks up our constantly sinking raised beds throughout the garden. Because of the peat bog ("you crazies," the old farmers say, sighing), we can't garden on the flat: ground-level beds turn into ponds at the first good rain. And so compost forms the bulk of our beds, sweetening the soil and helping the worms to open the tilth into something right for carrots and the small, questing snouts of snow peas. Every spring, the whole plot comes up in weeds, engaging us for weeks. And every summer, ouroboros-like, the garden consumes itself and emerges again, well fed and lush.

CHAPTER FIVE

The Memory Gardens

"You know what I think?" she says. "That people's memories are maybe the fuel they burn to stay alive."
– Haruki Murakami, *After Dark*

MINE IS A FAMILY OF GARDENERS. My British grandfather, sleeves rolled to the elbow, coaxing Peace roses out of the flinty Norfolk soil. My maternal grandmother twining whippet tendrils of star jasmine, bougainvillea. Even years later, in Toronto, having left Tanzania behind for good in the semi-formed East Indian exile following Independence, she always grew jasmine. My mother and I share this love of gardening, of flowers and scent. It's borne in the blood, frustrated constantly by Albertan weather. And of course, what did I do but complicate that horticultural urge further, moving to our northern farm just two scant growing zones off the Arctic.

I come to my passion honestly. My escape is into green: stitch an X into the soil, heel in something that will blossom, will cast its fragrance over what is left. The act of planting is a kind of alchemy, as though by squirrelling those snub-nosed bulbs into the earth every autumn, we can separate the roots from their baggage of beauty. Grief can be left behind, forgotten, when the green tops of daffodils nudge their way into late-spring light.

Everything lays down its burden of memory. I plant lavender in our farm garden for my British grandmother, its familiar camphor scent sketched into the skin of my hands. I know it won't survive this northern land but am willing to try and fail anyway. I plant roses for my grandfather. No Peace rose here, too much a prima donna, but a fistful of scruffy rugosas wild enough to hold their bristly leaves up against early frost or leafcutter bees. The fragrance of those rough blossoms, blowsy and unkempt, threatens to pull the garden apart at the seams every summer.

Batik-print irises spatter the spring flower beds, beloved by a good friend I haven't seen in years. White Nancy dead nettle stretches its thin runners in honour of a mentor of the same name, gone for a decade. Come midsummer, the sweet peas cast their shattering scent and the peonies hang in their cages, bursting petticoat blossoms in the June rain.

In the forest, out under the tall white spruce and navigable by a willow-chip path, there's a second memory garden composed of nothing but light. Here, after dusk, beeswax luminaria warm the dark, a handful of small harvest moons tucked away in the blackness under the trees.

Thomas and I count our dead in these gardens. They've become places to lay down their weight, places where we can go to remember, returning to teaching in the city relatively unencumbered. Grief loses its claws when there's a patch of ground in its name. I dig the

soil in our memory gardens and think of a quotation from friend and
fellow gardener Lisa Martin's first book of poetry, the beautiful and
wracking *One Crow Sorrow*:

That year, another failed miracle –
even the slim seeds she bothered planting
in her garden never came up:
she dug down looking.
       Remembered the names
and where she'd drowned them,
sugar pea, yellow bean.

Those small white flecks,
compost underground; they
fell apart at being touched.[1]

As the gardens offer us food and calm, they also offer us solace,
especially as we are two transatlantic hearts with our dead in many,
many grounds. In planting, there's a semblance of rebirth and an-
choring: *this* memory to *this* ground. And so there is the flower
garden out in the sun, brilliant and alive with bees, and the forest
garden, an unexpected calm space under the conifers. Eventually,
a bench by some wild columbines, antiques in outlandish bonnets.
The promise of bog orchids in summer, if the deer don't nip up their
June noses before they can flower. Twinflower looping its runners
over everything and blossoming in a froth of pinkish-white.

Like memory, the flower garden has stood the test of disaster:
floods of snowmelt, July downpours, hordes of grasshoppers. There's
no turning for help here; the garden must bear up on its own. We're
as fond of the overblown English flower gardens of my childhood
as any besotted green thumb, but they can be recreated here only in

cold-country moderation: Anchusa becomes catmint, sweet autumn clematis is replaced by golden hops and the towering trumpet lilies with their penetrating scent are swapped out for evening-scented stock, sweet peas, simple lily of the valley.

I can't remember who it was that said scent is the fastest conduit to memory, but I'd swear it's true. I can't count the number of times I've walked through a garden, smelled a certain fragrance and been carried off to some distant point in the past or to somebody's living room in a long-ago house. And thus, the tall white pipes of nicotiana, the musky scent of marigolds, the achingly sweet mignonette. Each chosen, each sown, to evoke someone no longer here.

They're interesting mediated spaces, the memory gardens and the market garden, suspended halfway between domesticated and wild. We plant with intention: kale and chard and spinach for salads and the freezer, beets and tomatoes to can, strawberries and rhubarb for jam and crumble. Potatoes to fuel us through long winter months of shovelling and snowshoeing and brush clearing, times when starchy foods are not so much a craving as a basic necessity. We sow to feed ourselves and our family through the three seasons when little grows outdoors. And we plant to remember the ones who aren't with us anymore; there's an aesthetic of loss in that. But we also plant for life and continuity. In our case, that's for the wildlife, in mimicry of what works outside the eight-foot moose fence ringing the garden. Milkweed and asters for the butterflies, daisies for the bees. Dill and chervil lift lacy white umbels like those of the cow parsley over the fence, beloved of tachinid flies. The garden is a curious interface between the wild world and the domesticated one, where the best of both brush up against each other and are selected and reselected over again for success.

ON A LATE October evening, the weak light in the farm garden finds little left above ground. A few plants of cilantro gone to seed, the dill drooping its massive heads and promising a haze of acid-green wildings at the earliest sign of May. They're here like memory over the winter, all the ones we have laid down, all the ones we still carry, uncertain, judging their weight. Out here under the snow, we hear them ticking over, murmuring us toward spring.

CHAPTER SIX

Cartography

LET ME WALK YOU THROUGH late June here in the North Country.

Start with shadows out on the township road that runs seven small hills down to our farm. Midsummer, just after the solstice, stand with your arms outflung on the highest hill, the sun wrapping your shoulders at nine p.m., the larks pinwheeling a final hour through the liquid air. To either side, long wild grasses going to flower and alsike clover fizzing with bees; beyond, the neighbour's canola crop ripening to a soundless gold. Other years, summer wheat, still young and blue during this month, heads shelved for August.

Walk with the sun behind you, the cooling land lifting a night breeze around your ankles, to the bottom of the hill. Static ping of grasshoppers off road crush, sweet wild alfalfa in shades of purple, poplar leaves just past sticky green, everything about them the scent of rising sap. A slight fog over the ground rolls as you walk: the willows have let fly their seeds this week, and you will be chasing them through the garden beds for the next month. This clock of seasons, of seeding: past willow, not yet balsam poplar.

The road ends where you can feel the night air pooling. One smaller branch turns off to the neighbour's driveway and his house overlooking the Paddle River; another continues onward through the trees to our farm. Well-met here at evening: the old grain bin that started it all, tin-flashed on a sagging wooden base, "Butler" in faded green paint across the panel above the door. The driveway is less a road than a hopeful track carved out of the dead brush at the edge of the property and scarcely holding its own against the trees. We have no illusions about how quickly the land will reclaim this farm should we move on. You might think it a constant threat, the push at our boundaries, but it's a relief, that knowing. We don't mind our tracks erased by green.

And erased they almost are, every second summer, if we don't top this wisp of a road with fresh gravel or drag it clean. The willow and poplar fluff take root and grow, the twinflower and wild strawberry are in perpetual creep, and everything is netted with a fine web of roots. Small plants slip under the fence that has always succeeded at keeping the moose out, and always failed spectacularly at repelling the nettles, and blend themselves in with the lawn grass and the patio stones. The boundary between wild and kept land is tenuous here, constantly shifting. That, more than anything, keeps us humble.

Seventy feet in from the township road, where the gravel track bends, the market garden opens out to the right. All told, the garden is about a hundred feet from the neighbour's fields, conventionally farmed, but this is a different world, especially in June. The tool shed and cookhouse, formerly the location of the old truck camper in which we spent our early summers, shine a vibrant blue in the summer evening. Against them rise cobalt delphiniums and Asiatic lilies in varying shades of yellow and burgundy, nearly ready to bloom. Golden hops scramble the moose fence alongside prairie traveller's joy clematis, sweeping from post to post and swathing the birdhouses in small pockets of shade and silence.

The delphiniums bleed a pool of blue onto the river-stone patio, our outdoor living room for three seasons of the year. Here, the cast-iron chiminea carries a fire most nights to heat water for the outdoor shower. In the early spring, it burns green wood to keep the mosquitoes away before the dragonflies hatch. We cook and eat under the sky, fireside, watching the light change on the treetops and listening to the sound of the forest around us shift. Bee hum to mosquito hum to dragonfly wings, bat clicks, pygmy owls. Coyotes. The progression toward nightfall is in my bones, just as we read the stars overhead in all seasons, knowing their expected passages.

The garden opens its long ranks of beds, food and flowers intermingled, herbs and sour cherries, zinnias and shelling peas. The greenhouse turns opaque at dusk, hatches battened against the cool air, keeping the peppers and tomatoes warm in case of a cold night. Our little frost hollow has its moods all months of the year, and no night is without danger. After much trial and error, I have long since given up on thermometers and forecasters. My own feet, bare on the cooling earth, tell me what I need to know of the night ahead and what it will bring this garden.

THE GRAVEL TRACK bends around the moose fence, and where it bends, the old grass seismic line runs off for a quarter-mile down to the hayfield. This is our temperamental route to the higher ground at the other end of the property: this bush road, picked out in boggy spots and unexpected spring swamps, pelting straight through the peat bog up to the hayfield. Someday, we'll drop lengths of dead tamarack and wedge them up tight, top them with clay or gravel to make ourselves a more reliable corduroy road, just as the highway crews once did through the bush along Highway 33, the Grizzly Trail itself. The first spine of that small highway, as with so many early roads across northern Canada, was a ragtag assemblage of trees, in slight but constant motion underfoot. Something about it appeals to me. I mourn the loss of that limber backbone, though I appreciate the springtime strength of blacktop and road crush when the meltwater comes.

The new pond lies outside the curve of the gravel track, its banks still unformed there at the edge of the forest. It's not yet deep enough to last through our intense summers, but it's proven itself a miracle of sound when the chorus frogs take their legs and leave the water. The pond is its own sort of instrument: wood ducks and Canadian toads, solitary sandpipers. Curtains of rain folding across its surface. Its rich presence hovers at the edge of my awareness, as water tends to do, wherever I am in the garden.

The cabin sits at the end of the gravel track, a simple pine structure, its back turned to the northwest wind and windowless. Its eastern side is mostly windowless, too, to afford a warm, strong flank indoors when the forest seems to be pressing in at all the other windows. The south and west sides are fully windowed, opening onto the deep woods and the garden. A small front porch, four feet deep, is just large enough to shelter drying firewood and give us a place to sit out of the rain. The cabin's main talent is to reveal itself as much bigger on the inside than it appears on the outside, a magic trick of light and space.

A summer or two from now, the gravel track will skirt the cabin and go on into the deep bush, circling through a clearing we're only now enlarging, where the birch and tamarack have died off due to time and porcupines. This will be where we raise the barn, a nod to the older farms that surround us, with space for the farm truck, a summer kitchen and a wash house on the ground level, as well as a study upstairs. Our farm is so far off the electric grid that power poles would be impossible, and so the barn roof will house a full solar array and a solar hot-water collector for the wash house. We've generated much of our own power on the farm from one small solar panel all these years; it will be an odd luxury to someday have enough power to run a yard light or a well pump.

Behind the garden fence, close to the barn, we have plans to expand our small field of heirloom wheat to include oats, barley and grain amaranth. Some of the additional dead wood will come down to give us more space for potatoes and pumpkins. This new earth will appear slowly, though, coming clear at a tree-by-tree pace. Thomas brings the snags down and bucks them, chopping the trees into moveable lengths, and I shift them, log by log, back to the cabin for splitting. After we've cleared a space of ground, we know it inti-mately: what kind of trees grew there, what underbrush still exists, the structure of the soil, the quality of light. By the time the main garden space had been opened, I had walked that ground thousands and thousands of times, pulling roots, hauling logs. I knew that earth as I knew my own skin. Even today, with a step on the ground or a handful of earth, I can feel what's going on with the soil in the gar-den. We turn to human power and a human time scale every chance we get, working with this land.

Someday, there will be a handful of river-stone patios between the barn and cabin and the trees. I imagine a couple of guest cabins for family when they visit, or for artist friends looking for a quiet place away from home and off the beaten path. Above all, we want this to be a place that is shared. We hope for more gatherings on this land, barbecues in the long-light evenings of July, harvest gatherings. Spring planting bees as the weather warms through May.

Beyond the cleared gardens, which extend at most two hundred feet from the grain bin, there's the forest. We've made walking paths here and there, but these are mostly used by the deer. The deep woods have their own character, rich and resinous during the long days, cool and intense at night. Whatever we're doing in the woods during the day, we leave by evening. Aside from the coyotes, the energy of the woods turns when the light wanes, heralded by a slight chill along our collars, and the forest becomes a place best left to itself.

THE FARM IS a place that always exists, as it does on this June evening, half in the real world and half in imagination. Back in 2006, when we drew up at the front gate and looked into the abandoned driveway, we could never have dreamed that this space would come to look as it does now. Even today, it's constantly shifting with hopes and plans and projects. We work with the land as spaces change: boggy areas grow, tree belts mature, old thickets dry out and die. I've mapped it for you as best I can this summer night, and even now, we're losing sight of the land's edges as the short northern dusk finally falls.

Halfway between the real world and the imagination isn't too bad a place to be. I suspect that all deeply felt projects and landscapes inhabit a similar space, revered and only semi-articulated. The land teaches us much about becoming comfortable with change, and with time, we learn that the best alterations of space come from careful negotiations between our own small hopes and the need to protect this place.

CHAPTER SEVEN

Cougar Country:
Living at the Edge of the Wild

IF I LIVED ANYWHERE ELSE for the sheer love of it, it would only be farther and farther north, chasing the boreal up to the Yukon or the Northwest Territories. There's something about living beside a great stretch of forest, both as participant and as witness, that is endlessly absorbing, at once enchanting and distressing. The former because there are vistas and qualities of light in the spaces of the everyday that are otherworldly, requiring an absolute halting of all activity and an undivided attention to just *that* light at *that* time. The latter because there is an incredible amount to learn to feel as though you have some small right to be here, holding fast on the patch of ground you stand on. Much of it comes from knowing the names and uses and history of the life around you. You cannot live on the land with any sort of consideration and awareness without being equally aware of all the spaces in which land is abused and mishandled. The province I hail from is the seat of Big Oil in Canada, the locus of disaster after disaster leaking like tailings-pond ichor out of the North. Time and again, our boreal is blasted by toxic spills and chewed up for open-pit mines, while farther south, we live the lie that land so badly mangled can be reclaimed in any true way.

Our farm hasn't quite escaped the long grasp of oil, either. The quarter-mile grass road linking the township road and the hayfield is a former seismic line, where a drilling rig slammed through untouched brush to test whether there was anything under our land worth taking. Luckily, the answer was no. The Bearpaw Shale is thick here, difficult to pierce even for a water well, and there is no oil or gas reserve underneath worth plundering. We've taken care over time to remove much of the evidence of the exploration, felling dead and half-chewed trees, allowing the verges to grow in and repopulate themselves with grasses and flowers, until what was once a wide belt of industrial carnage is now a simple grass walking path out to the hayfield shelterbelt and the wild saskatoon patch. But the intent behind the scar remains, and across this province, the boreal bears the brunt of our global obsession with easy oil.

WE ARE NOT the only ones running to this land to escape the damage of industry. Any large patch of unturned forest is almost unheard of in farming and ranching country; not until you reach beyond the Swan Hills farther north do you find the true untouched miles of muskeg bog that skein up toward the Arctic. A plot like ours is a haven for everything running from oil wells, fencelines and development. We don't have tigers, but we do have mountain lions and bears – *oh my*.

We get cougars on the land because it's wild old forest and a haven for deer and rabbits. When the sign goes up in Barrhead that there's a cougar in the area and residents need to watch their children, it's almost guaranteed that the cat will find its way out to our farm. Which is fine during the summer months when both Thomas and I are working together in the garden, and friends and family come and go: the place is busy and most big wildlife keeps its distance. The problems come in the shoulder season, when I'm out at the land alone.

A student once told me I was six inches shorter than my personality. That made me laugh, but the reality of it is that at five-foot-two, I look like *food* to any cougar so inclined, even though I make a lot of noise out on the farm. (So far, none of them have been inclined, but they do get awfully close to the cabin at times.) Winter is the best time for spotting signs that they've been near: we sometimes wake up after a fresh snowfall early in the season to find paw prints the size of my palm making a wide circuit of the woods. We'll find hunting tracks when we're out on our snowshoes in the deep bush.

Spring, though, is the worst time for a cougar to come through. It's early in their season after a hard winter, and I'm generally out there alone setting up the garden beds for planting (my teaching term ends two months before Thomas's does). Occasionally, a cougar will wander through around dusk, which in the North Country is about half past nine or ten. Spring is mating time for the wood frogs, so they're out in the pond and the shelterbelt in their thousands, singing their heads off. Then a cougar will pass, and I *know* it with every fibre of my being. The frogs will clam up for moose wandering nearby, and sometimes for humans, but for cougars, a band of silence develops that tracks the cougar as it walks. There is very little on earth that prickles the back of my neck quite like hearing that stretch of silence descend on the forest on a cool night in early spring. That's when I go inside, lock the cabin door and put on a roaring fire for tea to take the chill out of my bones. We've never had an incident, and I hope we never do. I won't take chances myself, and I still won't let friends with small children allow them to run out of sight on the forest walking trails. It's common sense and respect. If we're going to share this land, then we need to take precautions to allow that sharing to work – I'd much rather the cougar take a deer than a farmer.

Of course the cougars come through: the forest attracts more than just apex predators. We also have a healthy population of deer. One of our neighbours set up a motion-activated camera along a deer trail on our farm a few years back and found thirty coming through to bed down. And then there are the moose. It used to be a yearly autumn amusement, before we put up the eight-foot moose fence to stop the bull moose from sleeping on the potato plants, to sit by the outdoor fire in October and watch the bulls wander through the garden plot, calling plaintively for mates. We learned pretty quickly that although moose are enormously shy, they do become accustomed to our voices, so it's not unusual to find us working in the garden, talking quietly while one of the resident moose grazes willow shoots thirty feet away outside the fence.

For so many of the animals passing through the land, if they stick around long enough for us to get used to them, we're going to give them a name. The familiarity stops with that (we don't go for domestication), but it's nice to look up from weeding and recognize a moose or a deer by sight as it wanders by. We watched Spare Parts grow up from the world's most gangly calf into a stunning young bull moose. When autumn came, he moved on, following his instinct, but he's come through a couple of times in subsequent years. Bullseye probably ended up in a neighbour's freezer pretty quickly; although we don't hunt, we don't encourage the animals to come too close, as we want them wild and self-sustaining. But Bullseye figured that any human without a visible gun was a good human, and he'd stand out in the road a few metres away, staring calmly and gormlessly, a big hairy shooting-range target. Cold Shoulder shunned the bulls with aplomb for a couple of seasons until she got knocked up and came back one spring with a beautiful little calf; we caught them one morning sneaking onto the front porch to stare in the window. (A moose sneaking is a kind of thunder all its own.)

FOR ALL OUR living on the Grizzly Trail, that beautiful stretch of Highway 33 in northern Alberta that runs from Gunn up to Kinuso (just south of Lesser Slave Lake), there are no grizzlies. Although the Pembina and Paddle Rivers meander nearby with the slow brown eddies of most prairie waterways, their underbrush cover is thin. This land would be one heck of a walk for any grizzly inclined to head east from the Rockies, and there are no plains grizzlies left. But we do get the occasional black bear; they're large enough to startle and quiet enough to be completely unexpected when we're out in the forest during berry season.

Probably one of the hardest things about being a small farmer is the need to leave the farm for long periods of time to do two things: earn the money that keeps this developing farm going outside of summer vegetable and firewood sales, and teach and talk about the land with others. We love all we do, but leaving the land is always hard, even if it's coming into the city during the week to work in the classroom, the other thing we adore. My job has been more fickle than most: in the past, university sessional work meant a complete lack of job security and the potential to go semesters without work if classes were under-enrolled.

We were having "the talk" that many small farming couples have: whether we'd have enough money that month to afford necessities for our newly built farm, or whether we might actually have to leave the farm for good in order for me to get a full-time teaching job. It was the end of a very long summer's day of work and low spirits, of wondering how to hold it all together, wondering if we could. We'd gone out to the front gate, intending to take a walk down the township road after dark for some relief. I was feeling particularly low because the reality of it was that Thomas, in his mid-fifties, wanted to retire fairly soon after three decades of teaching, and on my uncertain sessional instructor's wage, there was no way I could support us both. It seemed one of us needed to be working full-time off the land so that we could survive the coming decades as we built the farm from the ground up. That one would have to be me.

I don't hold too much to talk of fate, but sometimes life confronts you with things that you just can't think your way around, and if you don't get the message, more fool you. Twice before in my life, I'd been told by Native elders I respected that I had an affinity for the bear, that it was my totem. I'd been told that, like the bear, I had the ability to go into dark places, starvation seasons in my life, and emerge again. Those words resonated. Thomas and I knew we had bears coming through the land; we'd find scat on the grass trail to the hayfield during berry time, and occasionally we'd wake up in the morning to the scent of bear musk outside the cabin. That evening, weighted down by even the notion of having to leave that land we loved and wanted to protect, we headed for the front gate and the moonlit stretch of township road. As I turned to close the gate, as clear as daylight, I knew something was out there in the bush. We shone a light into the trees, and the eyes that glinted back were too low to the ground for deer and too high for coyote. They watched us, motionless, for some time. And just as I breathed to Thomas, "That's a bear," the animal let out a low huff, the universal sound of a bear warning of its presence. We backed off in the dark and headed for the cabin, and it retreated into the trees. In the morning, there was no sign.

But the sign was plenty clear to me. This land is in my blood; it's who I am. Everything I do – writing, teaching young people how to write creatively and think critically about what we're doing to the landscapes we inhabit – finds its anchor in the land I am fortunate enough to call home. Moving away and selling the land would cause the end of so many things. At the core of it, I'd feel like I was losing a key sense of alignment that brought work, writing and daily life together.

I'd have to find another way to support us, but at least I had my answer.

Sometimes signs happen. Call them what you will . . . when they're as plain as your totem staring out at you with the dark weight of the forest behind it, you listen.

CHAPTER EIGHT

The Year of the Pond: Ecosystems and Adaptation

I think of the country as a kind of palimpsest scrawled over with the comings and goings of people, the erasure of time already in process even as the marks of passage are put down.
– Wendell Berry, *The Art of the Commonplace*

THE CANADIAN BOREAL IS A constantly changing landscape, reinscribed over the decades by ranching and farming, and most recently, by Big Oil. Over our lifetimes in this province, travelling either for pleasure or on earlier trips to look for our own land, Thomas and I have watched farms subsumed by industry as the big cities expand ever outward. We've witnessed thousands of acres go from good, arable land to suburban drive-only communities and RV storage lots; further out, we've seen old farmsteads turned over to gravel pits, pumpjacks and compressor stations along pipelines. We've wondered aloud about the ecosystems in those razed forests and drained sloughs that will never be replicated in the industrial lots or "lakeside living" communities mushrooming up in their wake.

Our small farm is surrounded by the last stand of agriculture before the true boreal – which in Alberta means muskeg spruce and pipelines – takes hold and stretches up to the Far North. Our neighbours run beef-cattle operations, dairies and large-scale farms. We watch from year to year as they're forced to clear more and more land to keep up with the costs of patented seed, chemical sprays and large machinery. Their farms are beautiful when the grass is in lush July growth and the cattle are out. Their places are absolutely bucolic when the wheat fields are ripening off in the September sun. But there was beauty there before the expansion of monocropped fields, too. The original sloughs and forests had their own cast of characters.

Long consideration and guilt went into the building of our market garden. We walked the quarter from end to end, trying to decide which place would be best for a garden and cabin; which would provide accessibility without altering too much of the landscape. Even our ultimate decision to take down the dead willow thicket nearest the township road and clear the land for our own little plot came after a great deal of pondering and regret. We're still trying to plant an equivalent number of trees elsewhere on the farm to the ones we had to take down.

The dead willow thicket had created its own small ecosystem each spring when it flooded: it was the frog breeding ground for this piece of land. When we started building the market garden, we heard right away how the fencelines and wild patches of brush resounded with frog song every spring, and we promised ourselves and the land that we would build a proper pond nearby.

It happens, when you need to work off the farm for money, that projects get bumped to the back burner, to other seasons. And so we found ourselves bringing in a small Bobcat in November so that Thomas could start shifting dirt for the pond out front of the cabin. For two full days in the failing light of early winter, and then again on a cold weekend in February, he worked to lift off the three feet of peat and dig down into the blue clay we'd discovered out in front of the cabin to build a watertight pond that would fill to the brim come spring runoff.

As with all of the projects on the farm, we'd started out with a pile of research. There's no mysterious guide to getting these things right, but there are stacks of great videos and magazines and books on modern homesteading out there. We read blogs about the lives of other homesteaders in Canada and the States. We watched YouTube videos on building ponds and creating ecosystems. We took a weekend course on building dugouts and small lakes from the hydrogeologist who'd come in during our first year on the land to assess the water table at our farm. We read constantly. And we talked to neighbours. In one brief chat, we learned that our closest neighbour had dug his own well and hit good water at seventy-five feet, far before the Bearpaw Shale could become an issue. Our problems definitely wouldn't include a dry pond.

After watching a documentary by Canadian environmentalist David Suzuki on the plight of Lake Winnipeg, whose waters are suffering paralyzing algal blooms because of high-nitrogen farming runoff and a breakdown of the marshes that once filtered the water coming into the lake, we decided to let the pond fill on its own in the spring to bring in a more balanced ecosystem with the natural water off the land. Instead of chlorinated water from the pumping station, we'd pile the pond full of snow and allow it to melt down, and encourage the wild water running in from the deep woods on the property to gather in the pond instead of doing its yearly overflow into the market garden. The wild water, natural surface water instead of chlorine-treated town water, would bring with it reed and bulrush seeds, not to mention all the tiny critters that inhabit a natural waterway, and the pond would establish itself as a working ecosystem in a short period of time.

We were captivated by so many things at the start of the project. We'd never dug a pond before, and where I was thinking of a small ten-by-ten-foot water garden, Thomas was excited to go all out and create a large dugout that would also provide a water source for the resident moose and deer. We were keen on the idea of a wild-water pond that would act as a sequestering place for anything that might wash in from the surrounding fields, much as the marshes around Lake Winnipeg, when at their best, act as giant systemic filters for the lake itself. The shelterbelt between our land and the neighbour's conventionally farmed fields is ditched on the far side to collect the runoff from his property and direct it away from our garden, but we were interested to see what the water quality in the pond would be, and whether being surrounded by Big Ag would have a visible impact.

And so we piled up the snowdrifts in the pond to wait for spring. But remember that bit earlier about how working off the farm changes the task list quite substantially from season to season? This proved to be one of those occasions when working off the farm as teachers turned around and bit us: after a February weekend in the Bobcat, finishing up banking the pond and piling the snow, Thomas slithered the machine out of the clay and onto the road for us to clean off before taking it back to the rental place in town. The peat and clay, however, had other ideas. While Thomas was working, the frozen dirt had been flying. Unbeknownst to us, it was thawing where it hit the hot machine and then freezing again. The minute Thomas hopped out of the machine and killed the engine, the rapidly freezing dirt locked the tracks into place.

We'd hoped to dig the pond during the fall, when the ground was dry and the temperatures were kind, but there we were instead, blowing into our cupped hands to keep them from freezing in the minus-twenty-eight-degrees-Celsius evening as we chiseled a shell of frozen peat off the Bobcat. It was a Sunday night, we both had early classes in the morning one hundred fifty kilometres away, and the Bobcat was too frozen to drive up onto the trailer. I chiseled away with a screwdriver for two hours while Thomas moved the tracks back and forth to keep the dirt I was loosening from freezing back in place. It was solidly dark by the time the treads came free and we could load up the machine, its complaining engine ripping through the still February night. In the end, nothing saved us but an hour in the truck wash in Barrhead, blasting the peat off with a high-powered sprayer that'd strip your skin off if you misfired. By the time we got the machine back to Edmonton, it was coming up on midnight and the damn thing had frozen solidly to the trailer (the guys at the rental place sure had fun with that the next day!). But in the end, like so many mad farm projects, the pond proved itself incredibly worth the adventure.

Saskatchewan poet Barbara Langhorst speaks of May floods on the Prairies and "a house hoisted on pools of eyes/silver duct tape shimmer/frog song seep[ing] in at the sills."[1] This is how spring came to the farm in the Year of the Pond. What had been an unearthly moonscape in November and an ugly bomb crater in February now filled itself to the brim with clear meltwater and, by some sort of arcane wild-water magic, hundreds of defrosted wood frogs. The nights were loud with their trilling, growing silent at the approach of anything larger than a coyote, and the days saw the new pond fill with thousands of tadpoles. Animals came down to drink; dragonflies and bats circled overhead to feed on the small insects rising from the surface. Ducks and sandpipers moved in, and the moose occasionally wandered through to eat the bulrushes. The pond easily raised a couple thousand wood frogs, which left the water in a mass exodus in July and entered the garden, where they spent the summer keeping the pests down and dodging the lawnmower. I was forever moving handfuls of tiny frogs out of the pathways when weeding the rows of growing vegetables.

But, like all good things, the pond came to an end at the start of September. Occasional saturating rainfalls weren't enough to combat an overall dry summer, and by the start of school, the pond was dry. By then, the frogs had all moved out and the resident sandpipers had raised their young, so the loss of this small ecosystem wasn't as much of a shock as it might have been back in June. It was a warning, though, that although we got a lot right, there was still a long way to go in creating a self-sustaining year-round ecosystem. Pond success starts with digging deep enough, and it includes banking the sides tightly with clay so the water isn't likely to seep out into the surrounding soil. The past few summers at the farm have been wet ones, and even the neighbouring dugout built on sand has been staying full through to autumn instead of leaching dry as it often does in drought years. But there's no shortcutting here: we want a pond that will last, whatever the summer brings.

We're already eyeing the end of this summer as a chance to get out in the Bobcat again, while the ground is dry and the weather is holding, and dig the pond another ten feet deeper into the heart of the blue clay. That should hold the water year-round and allow us to stock the pond with a solar-powered bubbler and some fish. Maybe I'm idealistic, thinking that at some point I might be able to trot out the front door with my fishing rod and bring in a small trout for supper, but if we weren't idealists, we wouldn't be out here on our little farm in the woods in the first place. And the very best part about it all is that nobody holds us back and says, "Not in my backyard!" At the same time, though, if we do something wrong and end up harming the land, the burden of that lies on our shoulders, too. It's both a joy and a responsibility, working with any wild place.

Ecosystems can be created and destroyed over a very short time: while we were digging our pond, the neighbour down the range road was pulping an entire forest that used to be alive with moose and deer and birds. Over the space of two weeks in late winter, he brought in an industrial brush mower and a bulldozer and flattened a quarter section of thriving poplar and spruce. When we drove across the Paddle River bridge and swung round the corner, the sight of that shredded forest brought me to tears. Everything in the boreal world exists in a fine balance, and when we work off the farm, we're not always as closely in rhythm with that balance as we'd like. But we are always aware of it and deeply respectful of it, and being able to build an ecosystem and watch it come to life over the course of a summer has been a privilege indeed.

In the meantime, I'll stash my fishing rod and my lazy-afternoon-rowboat dreams away for just a little longer.

CHAPTER NINE

Bring on the Bees

The keeping of bees is like the direction of sunbeams.
– Henry David Thoreau

IT'S EARLY SPRING, so early that the snow still hasn't left the ground out in bright sun. It's the first day that temperatures have warmed enough to allow me outside without full winter gear, and I'm in a camping chair in front of the cabin, surrounded by fairly high snow-drifts. First sunning day of the season: this means a down-filled sleeping bag to ward off the chill from the ground below and an old copy of *Mother Earth News* in hand, my eyes narrowed against the glare of sun on snow. The heat is picking up – already, in early March, it's got the same softness to it as a warm October day – and my bones welcome the heat and light after a dark North Country winter.

Plink. Something bops off the magazine and falls to the ground. *Plink.* Behind me, something else collides with the cabin wall. Suddenly, I'm aware of a constant, gentle sound all around me, and I realize that it's *honeybees*, winter-dizzy and on their first flight of the season, crashing into everything and sundry. They're tiddly on bee candy or sugar water after a long winter, and like me, they've been called outside by the sun before the season has quite broken.

The first bees of spring hold an uncanny fascination for me. It's as though they've popped out of a snowbank, so suddenly and unexpectedly do they appear. Some years, it's ages between their first flight and the flowering of the basket willow, and I wonder what on earth could be sustaining those minute golden darts while the rest of the landscape catches up with them. I sow bee-friendly flowers and trim the willow to encourage fresh catkins, but somehow I never seem mentally prepared for that first tipsy spring onslaught.

OUR MARKET GARDEN has always been for the bees. The planning behind it goes something along the lines of *a row for me, a row for the bees*, so closely do we interplant vegetables, fruits and herbs with flowers. Delphinium does double duty, beloved by hummingbirds and the big solitary bumblebees in their improbable flight. Saskatoon, haskap and sour cherry break into a froth of early blossom along with flowering plum and bearded iris. Rhubarb is allowed to go to seed in spectacular plumes of greenish flowers because the panicles swarm with bees, and lines of cosmos and pinks buzz through the summer months. In the greenhouse, I let sweet basil run wild and burst into the small white flowers that, with oregano, draw tiny bumblebees by their hundreds. Once in the greenhouse, these little bees usually dawdle amongst the peppers and tomatoes and do some overtime there, emerging completely covered in neon-yellow tomato pollen.

We've always wanted to document the bees in the garden and have never quite gotten around to it, although every summer, plenty of time that might otherwise go to weeding is spent watching leaf-cutter bees snip segments from the leaves of the Hansa rose and marvelling at the bright pink "landing lines" that guide the bees into the hearts of blueweed blossoms. We've installed nesting boxes for solitary mason bees and observed, perplexed, as a gathering of large bumblebees took place inside the back of the camper fridge one autumn. One by one, they made their slow way into the heart of the pink insulation that wrapped the fridge and nested there for warmth over the winter. For the longest time, if we went into the drawer under the fridge to fetch a frying pan and returned it later with too loud a slam, we'd be greeted by the dozy murmur of the entire colony being jolted from slumber.

But we wanted honeybees. After many years of watching the native bees go about their business, we were firmly hooked, and we wanted to try our luck with a small colony or two. We took a course from an experienced beekeeper at the local botanical gardens, read endlessly, visited the city's bee-supply shop and planned to place an order in the spring. But just as the new growing season began and the willows broke, our hopes for a summer colony were dashed ... the orders came in, and the supply of new bees and queens was sold out.

OUR LOSS WAS partly from timing and partly from an increase in the number of orders. It had been a bad winter, long and cold, and many apiaries had experienced up to forty percent losses in their operations. We'd missed the boat that year, but we'd be ready the next with a mid-winter order. In the meantime, the extra months would allow us to choose the best possible site for the hives: the end of the garden in the small fruit orchard, where the bees would have access to cover-crop flowers and dandelions, the flowers of the small fruit shrubs and the surrounding thicket of basket willow.

We stayed that way for several months – in a holding pattern, waiting for bees. Thomas was excited by the notion of farm-raised honey somewhere down the line, and I was intensely curious about the whole process of extracting beeswax following the honey harvest. Our cabin is off the power grid, and during the long, dark winter months, we tear through our homemade beeswax tapers almost faster than I can keep up. Closing the loop on yet another element of farm life by pouring tapers from our own beeswax would bring us one step closer to the self-sufficiency we dream of.

My deepest reason for beekeeping, however, has less to do with products and more to do with process. I like the way beekeeping requires you to slow down, to be calm, to be kind. I've watched an agitated or threatened hive, and the smallest thing will set the colony off. I've also seen a queenless hive in a state of utter panic, the workers sensing that something is not right and passing the signal around until every last bee is vibrating with fear. I come to beekeeping enamoured of the way in which it asks you to slow your frenetic everyday pace and be aware of the hive as a living entity, a colony with moods of its own. I'm sold on the idea of calm, of matching your own energy to the bees'.

And so we waited out one last winter before the order of New Zealand bees came in. We spent our time reading everything we could lay hands on about keeping hives, dreaming about the plot of buckwheat we'd plant to surround the bees with rich white blossoms come June. In the meantime, we planned the rest of the garden as much for the local bees as for us humans, trying to have something blossoming in nearly every season, and especially the shoulder times of spring and autumn. We found Thoreau's words coming unbidden to mind as we pored over seed catalogues and books on apiculture, wondering what he meant, guessing that it had to do with the way a hive is a creature of its own, with a heat and a life force unique from its parts. The same way sunlight can't be directed or constrained, but is anticipated and reveled in, and the same way I will always find a way to be outdoors with the sun on my face when the bees come alive in early spring.

CHAPTER TEN

Growing Food in Frigid Weather

THE FIRST YEAR WE PLANTED an entire garden at the farm, Thomas was teaching full-time for the public school system and I was on a year's break, leaving behind a full-time career teaching high school and about to start another one at university. We'd inherited a couple of enormous homemade grow beds with light ballasts from a family member and had set them up in the living room of our apartment in Edmonton; even now, these same grow beds continue to start all of the frost-tender crops we plant out at the farm each summer.

So there we were, around the end of April, with a living room wall-to-wall full of these gigantic grow beds and more packets of seed than we knew what to do with. We'd begun gardening on the land with some potato beds the year before, and they'd done so well in the peaty soil that we were gung-ho to try everything all at once. We knew better (in theory), but the catalogues were so tempting and the winter had been so very long…come spring, we wanted to see as much green as possible.

All I can think is that we must have appeared to be the world's biggest and dumbest grow op to all the owners of the surrounding condos. From their living rooms, they looked straight into ours, where rows of tomato plants grew from seed under natural light during the day and fluorescent lighting at night. The grow beds are two-tiered, and that first year, we made the most of every square inch of illuminated space: you could barely navigate your way through the apartment's main room without tripping over a flat of onions or a tray of peppers. Our elderly building manager, whose preferred pastime was to sit out on his balcony in the summers and indulge in some deafening Sinatra and a home-rolled joint or two, thought he'd entered the promised land when he came up one day to do an inspection and found our front room full of leafy greens. He was crestfallen when we explained that they were tomato seedlings for our farm and that he could take one home to grow on his balcony if he liked. He allowed us to keep the grow beds up, but we always felt as though we'd slipped a little in his esteem.

OUR APARTMENT WAS located on the third floor of a walk-up in Edmonton's artsy Old Strathcona neighbourhood, and our manager likely wouldn't have had to go more than a block in any direction to find an operation that was selling his sort of greenery. It was a tired old building, subject to frequent and half-assed makeovers by the various owners; but where the carpet and siding could be changed without much issue, the guts of the old building remained a perennial pain in the butt for two would-be farmers. The radiators were senior citizens in their own right and absolutely refused to circulate hot water to the third floor where we, and our tomato grow op, resided. There were several cold spring mornings when we'd wake to the apartment holding steady at a chilly four degrees Celsius and scramble to get the oven turned on and its door open in time to stop the tomatoes from getting frostbite.

When I moved out to the farm in May with two hundred tomato seedlings and my parents' geriatric dog (ostensibly for cougar protection, but more realistically as bait), the weather at night was hovering just above freezing. For the entire month of May, Penny and I lived in that tin can of a truck camper, shivering by the outdoor fire at night, showering outdoors under a spruce tree (well, I did, anyway) and waiting for the temperatures to warm sufficiently for me to plant out my carefully nurtured city seedlings. Because let's face it: if eight months of winter, dipping to an attitude-slaying two weeks of minus fifty degrees Celsius at its worst point, isn't bad enough, the springs and falls on the Prairies are mercurial to say the least. It can be plus twenty degrees one day and snowing the next in September; equally, it can be minus fifteen degrees one day and plus twenty-five the next in March. There is little rhyme or reason to shoulder-season weather in this landscape.

All my childhood in Edmonton, I'd had the May long weekend ground into me as the safe bet for planting the summer's vegetables and flowers. Come the twenty-fourth and twenty-fifth, my mom would take the three of us kids out to Hole's Greenhouses and Gardens in St. Albert, and we'd compete with the other like-minded gardeners for starter plants and hanging baskets – all the things that Mom didn't have the time to begin at home with a plethora of kids around. But nothing, repeat, nothing ever went into the ground until Victoria Day was firmly behind us.

THERE WERE NO such rules at the farm. Some years, the temperamental spring would grace us with its presence as early as April; other years, we were still getting night frosts in June. There were three things you could do: read the almanac religiously, dare to trust the local forecasters, or grow plenty extra and hedge your bets.

Spring was cantankerous that first year, and so Penny and I got the seed potatoes in and heaped up the garden beds in preparation for planting, but held back on the tomatoes until one day in early July. It was over a month since Edmonton, an hour and a half to the southeast, had gotten its last night frost. My parents' city garden was thrumming with bees and growing like mad. But in the North, the open skies at night still brought ground frosts with them, and our farm showed its Achilles heel: its location in a frost hollow at the foot of the surrounding hills. Just up the range road, the night might be cool and still, but in our frigid little garden, the collecting cold air dusted the beds with frost crystals.

In early July, feeling we were safe, Penny and I finally set out eighty-five leggy young tomato starts, the same ones that our building manager had pined over in April. The days had been in the high twenties, and we were feeling hopeful: how late into the summer could frost arrive, anyway? I should have read the sky better and watched with trepidation as the wind pushed the clouds aside; as it was, I soaked up the sun on my back as I happily planted the tomatoes out, tucking a few away in the greenhouse for careful measure.

Dusk fell and the wind dropped, and my heart with it. Right away, I could feel the cool in the air and the damp slipping down into our bowl of a garden from the surrounding fields. I gathered together every blanket, tarp and sheet I could find and spent midnight with an industrial flashlight, tucking in the plants against what I now knew would be a sharp frost. But there was nothing to be done. When our garden frosts, and bear in mind that peat freezes like cement, it frosts completely. Nothing is left untouched. By the morning, the rising sun and the breeze woke me with the rank scent of frost-killed tomatoes. We lost all eighty-five plants.

This is partly the reason for this book: nobody tells you when you decide to start growing in a cold climate like ours just how tough a journey it will be. The learning curve is an overhang in locations with eight-month winters and worse. The diet of gardening books Thomas and I grew up on didn't include much about Zone 2 or ground frosts in July, and the folks who talk to us about growing year-round in cold climates by using extra polytunnels in an unheated greenhouse don't understand what minus forty Celsius plus wind chill for two weeks does to plant tissue, let alone human tissue. We take a perverse sort of pleasure here when our winter temperatures are worse than those of Siberia, and our gardening is a world of its own, very much a world apart.

That doesn't mean we shouldn't be doing it. We absolutely should – and the more northern the location, the better off we have it in some select ways. Our summer days are incredibly long, with light lingering in the sky at our farm until midnight around the solstice, and the sun starting to rise already by four a.m. Those long days make up for our shaky starts to the growing year, and plants catch up quickly to those sown in the early spring in other parts of the world. What we face here is the heartbreak associated with trying to grow crops in a region that is unpredictable. It's not unusual for us to lose the entire garden to frosts in mid-August if we haven't gotten the horticultural fleece out in time, and that first July frost wasn't our last. What we gain from trying to grow in a climate like this is the ability to hold on, hold out. Just because the sweet corn got frosted unexpectedly a few days before harvest one year doesn't mean we won't catch it in time the next. Just because one variety of tomato dropped its flowers all summer long in the cool and the rain doesn't mean that we can't find a better heirloom variety grown for colder areas like ours. It's true that not every sort of seed I throw at the ground will grow, or will really even have a shot during summers that can last from two months to five. But we have to try. Particularly during a time in which global food sources are under threat – more people on less land, and the land we have available increasingly polluted by the chemicals required to grow many GM crops – we need to try our best in cold climates to see what we can harvest for ourselves.

We're getting better at recognizing the signs of extreme weather on the farm, but the very real truth is that we lose crops every single year to late frosts or unexpected snows, not to mention garden pests and the occasional hailstorm. What we gain is the knowledge that the seed we keep comes from crops attuned to our location and chosen specifically to thrive and ripen in our frost-hollow garden before the nights get too cold. During the long winter months, if I've planted and harvested the garden well, I can sidestep the grocery aisles of vegetables flown in from California or abroad and instead turn to what I've canned, frozen or stored in the pantry to make the majority of meals we eat. It *is* possible to eat something from a cold-climate garden every day of the year and not have to turn to those underripe hockey pucks shipped from south of the border that are passed off to us as produce during our long Canadian winters.

CHAPTER ELEVEN

A Year of Farm Food: Community-Supported Agriculture

NOT TOO LONG AFTER we'd planted our first large garden at the farm, we read Alisa Smith and J.B. MacKinnon's *The 100-Mile Diet: A Year of Local Eating*. At that point, we were already eating, during the summers, from what we called "The 100-Foot Diet" – aside from meat, milk, butter and tea, we harvested just about all of our meals from the garden where we spent our days, cooking and eating in sight of the beds our food had come from. Increasingly, our friends in the city wanted to do the same, and realizing we were within reasonable driving distance of Edmonton, we ran with their enthusiasm and began a community-supported agriculture (CSA) food box program.

These programs have been around in one way or another for as long as farmers have been providing families in town with vegetables, eggs, milk and grain, but they've gained considerable cachet over the past fifteen years. When we began Larch Grove Farm's CSA program, we were feeling our way forward based on what we'd read and learned about from small farmers in Alberta and British Columbia. There were a handful of other small farms around Edmonton offering CSA shares at the time, and these were relatively limited compared to the number of people who were interested in joining the programs. Because this would be our first year as a CSA farm and we were farther north than many of the others, we played a very conservative game with our pricing and offered a ten-week box program in sizes varying from a personal snack box to a large family hamper.

The main thing to understand with a CSA program is that it's built on faith in both farmer and land. Customers pay up front for a summer's worth of food months before they actually see the first bunch of spring onions or bag of spinach. The money comes in around seed-ordering time, mid-winter, right when small farmers are dreaming of delicious crops and comparing catalogues, calculating over and over again what they can afford to sow in a given year. The CSA money goes into start-up: seeds, pots, tools and the like. It's just what's needed to kick-start a market garden in the late winter.

After that, the contents of a CSA box are up to the farmer, the elements and the pests. Most farms will list the potential contents of a produce box at a number of points throughout the summer, but this can change with crop failure (and, likewise, with spectacular success – you can always tell if a farmer's had a good kale year!). It may be variable in type, but the food is local, fresh and, with many small farms, grown without the use of pesticides and chemicals. For ten to fifteen weeks out of the year, a family with a CSA share can count on having enough weekly vegetables to satisfy the majority of their needs.

We began our CSA program on request from friends after they saw the contents of the vegetable boxes we were bringing back into the city for my parents, who helped us routinely on the farm through the summer. Suddenly, over the space of about a week's sign-up the winter before our CSA program began, we had twelve more families to plan for. We scoured catalogues and polled our friends: what did they and their kids most like to eat? What did they want to avoid? How keen were they to see something new and unusual in their boxes each season? We ordered accordingly, and through January and February, more seed than we'd ever owned before began arriving at the house. We ordered pots and trays, set up the trusty old grow beds and kicked production into high gear.

All I can say is thank goodness for my parents during those early years. Without their basement for the grow beds as our operation expanded and expanded, we probably would've blown out the wiring in our old apartment. Through March and April, onions and leeks sprouted and tomato seedlings popped to life in my parents' cool concrete basement. When the time came to plant in early summer, the laundry room was overflowing with starters. We spent every spare evening and weekend at the farm, working dawn until dusk to get larger beds tilled and composted. Our bodies ached and we were already thoroughly sunburned and well muscled by the start of June, but darn it, we were ready.

It's fortunate that we have always loved the work of the farm, or that first summer of CSA would've done us in for certain. We had calculated and planned meticulously through the winter: ideal sowing dates, transplanting dates, crop-rotation schedules, harvest schedules. We'd calculated how many rows of which crops we needed throughout the summer to maintain a good variety in the boxes each week. But we *hadn't* figured on the weather going off the rails. Nature did its own sweet thing.

AND THIS IS where I say to climate-change naysayers, just ask a farmer. More to the point, ask a northern farmer. The elderly couple from whom we'd bought the farm used to tell us that you could set your planting schedule in advance year on year and pretty much know the rainy months and the sunny ones; the shoulder season could go either way. But the year we began our CSA program, and the two summers after that, the weather went haywire. For weeks, it rained and rained and rained. June passed, and we planted our mainstay crops into the boggy beds because we had to keep to our harvesting schedule. July arrived, the rain finally drew to a close, and then an uninterrupted drought hit for the rest of the summer. The crops didn't know what to do with themselves. The greens drew in all that moisture and went bonkers; we harvested heritage butter-head lettuces so big that one plant looked like a bouquet in Thomas's arms. The Bloomsdale spinach turned out palm-sized leaves, and the radishes were the size of hand grenades (my closest friend com-mented on one of our crop photos and asked whether we might be farming a nuclear waste site, what with all the gargantuan veg-etables!). The more fair-weather vegetables, like the onions, sulked in soil that had been damp for so many weeks into the growing season. The runner beans finally outgrew their cold-soil start and flowered, just in time to be sniped by the first frost at the beginning of September.

In spite of the weather calamity, each week the garden produced beautiful (if unpredictable) loads of vegetables for fifteen families (our twelve boxers, plus my parents, some of Thomas's family and us). We harvested delicious young zucchinis in green and yellow, field cucumbers, carrots and cut-and-come-again greens. Our home-built greenhouse kept off the rain and yielded pails of tomatoes and Gypsy peppers. The bush beans managed a good crop, though the runners were a loss, and we had sugar snap peas, new potatoes, beets and more. Each Sunday morning, we got up in the five-a.m. pre-dawn light, then picked and packed the boxes we'd washed the night before. We rinsed and spun and bagged spinach right there in the garden. We sluiced and bundled carrots. Every week, we prepared in advance a recipe for the crops coming to their peak and topped off the latest harvest with a copy of the recipe in each box. We kept a farm blog with further recipes, details about the farm and our growing season, and stories of the crazy adventures we were living out there in the bush.

THAT YEAR WAS probably the most fun we've ever had in a summer. The twelve families signed up for our CSA boxes were all close friends, either work colleagues or part of our city community, and the food box program kept us linked over the summer. Each Sunday, as we drove the boxes into Edmonton and dropped them round at our friends' houses, we caught up on the news of their lives and their families. If we had time at the end of our route, we'd stop on someone's veranda for a cup of tea. The entire job took us nine hours each Sunday. From the time we woke at five a.m. and hauled ourselves out to work in the misty garden to the time we rolled away from the final delivery around two p.m., we worked non-stop. It was the single most enjoyable and tiring summer of our lives.

For two full-time teachers (back in the early days of the farm, I was teaching in the public schools too and hadn't yet made the switch to university), running a CSA program during the summer was a challenge. We began our spring deliveries a week before the teaching year ended in June, worked straight through the summer, and dropped off our final boxes a week after the start of the next school year. But there was something about being able to provide food for all those families just from our own labour in the market garden that was intensely rewarding. Our friends began to take a stake in the farm, and in August, everyone trekked out to the land for our first Open Farm Day, a chance to explore our farm, pick food from the garden and eat a farm-inspired barbecue around a wood fire with like-minded people. Kids roamed through the market garden, popping peas from pods and pulling each other around in the yard cart. Families began to grow interested in the heritage open-pollinated seed varieties we were sowing, and we enjoyed sharing the colourful histories of the plants in our garden. Looking around the market garden at our friends taking in the beautiful Alberta summer day, we found Walt Whitman coming unbidden to mind: "Happiness [...] not in another place but this place, not for another hour but this hour."[1] The CSA program brought all of us involved in it closer to the land and made us more aware of the seasonal rituals associated with our food.

We ran the food box deliveries for two good (though climatically unpredictable) summers, but in the third year of the program, right about the time we'd planned to sign up new shares, the farm received record snows. By March, we had four feet on the ground and more coming. In April, with the farm half snowfield and half lake, we had to concede defeat and suspend our program. The farm remained underwater through to the middle of August. The now-twenty families of colleagues and friends were disappointed, but they understood.

The wet summer was a crushing blow not only for us as new small farmers trying to grow a CSA program, but also for Barrhead County. Just short years before, our county had been declared a disaster zone for grasshopper damage; that year, it was declared a disaster zone for flood damage. After the heavy snowmelt, the Paddle and Pembina Rivers overflowed their banks, and suddenly, more than eight thousand acres in the county were underwater. It took us clear through the summer of 2014 to retill the entire garden and raise up the beds with fresh soil brought in from the pond excavation to make up for the soil that had washed away or gone anaerobic under-water, starved of oxygen and now sterile.

After another couple of years of intense cover cropping to rebuild the soil's fertility, we'll be on our way to opening the CSA program again. Thomas is looking forward to retirement, tending the bees and raising a flock of laying hens for the farm CSA shares, and I have plans to become certified in clinical herbalism and build a solid farm kitchen so that I can offer jams and herbal blends. We're lucky to have a patient waiting list of folks who are willing to navigate the hazards of a changing climate with us in the interest of good local food, sustainably grown, and the sense of connection to a place that comes from knowing their farmers like family.

CHAPTER TWELVE

The Cabin Comes Home to Roost

WHEN YOU BEGIN TO SPEND nights on the land, instead of just your working days, your whole relationship to the land changes. Ultimately, it becomes less about successful completion of the moment's task and more about bringing the body and the breath home to a sense of place. In a search to explain spiritual disorder in North America and point to what might reconnect the spirit to something larger than itself, Wendell Berry speaks of how he spent a great deal of time in his life "groping for connections – between the spirit and the body, the body and other bodies, the body and the earth."[1] I don't think it's too grand a notion that one small piece of land, whether in the city or out, might serve to connect one to a sense of something larger, and in doing so – in establishing that essential balance again – might help one to better connect to other people.

Our farm has been, from its very early days, a place of gathering: family members coming out to help with various garden tasks, friends driving from the city to walk the forest trails, neighbours dropping by to chat about the growing season. Early on, though, our way of knowing the land was limited because it ended at dusk each day. There are a slew of hours when the boreal forest becomes a different place altogether: the dark hours, when such things as passing cougars become very real. If you can't sit out in safety and hear the local coyote pack denning after a night hunt, or the neighbours a mile away sitting creekside by the firepit, you lose an element of place. Being in a space into the dark hours, watching how the sun progresses from season to season, anchors you. It's akin to watching the light change on a patch of ground you're just coming to know. Before any deep understanding can settle in, you watch that sideways slide of light through all the months of the year. You wait.

Michael Pollan talks about the concept of reduction, of "*simplification* – of reducing so many daunting new complexities to something as stripped-down and uncomplicated as a hut in the woods [. . .] a room of my own [. . .] a room of my own making."[2] This simplification was exactly what we'd been dreaming of when we moved onto the farm during the summer of 2007: a space of our own where we could pass all the hours of the day in learning the land; a space, we hoped, we could build ourselves.

It's no secret that Thomas and I had a lot to learn, coming as we did from the city to the farm. We're no different from any other family that makes the leap from one place to another: at the same time that we're learning a new space, we're also learning what it will ask of us, the skills and privations. We can frame a shed pretty decently, and we work well as a partnership on most of our building projects, but neither of us is a carpenter. When we decided we wanted to live on the land in all seasons, there was no bubble to burst. We realized early on that if we tried, like Pollan, to build a room of our own, it'd be an ugly, utilitarian box.

That's where Mark came in and saved our homesteading bacon. He was a woodworker and carpenter looking to grow his business, and we were looking for somebody with more skills than the two of us to craft a cabin that would hold up to our frigid North Country winters. In eight months, he'd built us a superinsulated sixteen-by-twenty-eight-foot cabin, tin-roofed so we could harvest rainwater, and wired to a low-voltage electrical system so we could run the cabin off the solar panel we'd used in the truck camper during our first years on the farm. While we were learning how to manage a Bobcat and move two dump trucks' worth of gravel to build a cabin pad and access road where before there had been only grass, Mark was putting the finishing touches on the wainscotting before he loaded the entire thing onto his trailer and drove it three hours north.

Sure, we had a few friends who thought we were ducking the full back-to-the-land experience by hiring a builder, but one thing we've embraced as modern-day homesteaders is the concept of going with the times and recognizing our limitations. Much as I respect the lifestyle led by homesteading greats such as Helen and Scott Nearing and consider them astonishingly tough, there's no way I'm going to carve my own eating utensils. Even Thoreau tossed the bean fields for town, which was, incidentally, nearby. We came out here for the land, and being on the land as much as possible is our primary focus. Hiring Mark to build the cabin allowed us to be on the land more quickly, and honestly, it was safer than the two of us trying to hack together a cabin on our own. It's been the hardest lesson to learn, but a necessary one: we don't have to live in a vacuum. We can hire Mark to build our cabin and still grow all our own food to remain self-sufficient throughout the winter. We can sell a load of birch to friends for enough money to keep the Jetta in gas for a month, and we can give up the landline in favour of a pair of cellphones that keep us safely attached to the outside world while also allowing our cabin to remain off the grid. Homesteading doesn't have to be a solo game: it's as interconnected as you want.

So the cabin arrived at our farm on an early October day in 2011. We watched the trailer approach down the township road like something out of a hallucinatory Prairie Gothic movie, surreal under a sky strung with departing geese. A few hours later and one tree less, Mark's friends had backed in the trailer, lowered the bed and hauled the cabin off using a large Bobcat and a lot of shouted encouragement. We scaled the roof peak to install the chimney together, stepped back and, as simple as that, the day opened itself like a book. Once the stove was in, we settled into the farm in a way we'd never experienced before. No longer were the long work days bookended with hours of road travel; no longer were we cramped into a tiny, unheated truck camper for months on end. At just two hundred fifty square feet plus loft, our little cabin gave us the ability to know the land at all times, in all weather.

Very quickly, the cabin became the center of our life on the farm. My days of shivering through early spring alone in the truck camper, setting up and planting the market garden while Thomas finished his teaching year in the city, were done. I could live solo at the farm in comfort and safety. I could watch the light change in the garden in a way I'd never known before. Working with the land from dawn to dusk, sleeping in that small, safe space next to the forest and the garden through the short northern summer nights, I came to understand that land to a depth I'd never done before.

When Thomas and I speak to people about the farm, we often hear in response, "Oh, it must be nice to have good health and youth on your side. I'd love to do something like that, but at my age, it's impossible." Over the years, though, we've come to subscribe to

Wendell Berry's view that if you can get out to the land in some way, any way, it's not all that difficult to bring body and spirit back together again.

THOMAS TOOK UP small-scale farming in his late forties, afflicted by incredible chronic back pain from a downhill-skiing injury sustained years before. Initially, he was worried that the land would ask more of him physically than he could give. The gentle work rhythm of the farm, though – its seasonal patterns and varied demands – has kept him so limber that the back pain is now a thing of the past. It's the ultimate gym subscription: the exercise we need is built into the natural rhythm of our days, and tacking on a workout at the end of a busy stint in the garden has become redundant.

At thirty-four, I already carry a history of hurt in my body: twice-fractured collarbone, disarticulated finger joints, several staved-in ribs. In 2012, unexpectedly hospitalized for emergency surgery, I came within a hair's breadth of dying. As the specialist surgeon confessed to me later, if I had been anywhere else than in Edmonton that day, I wouldn't have survived. Had I been travelling, he told me, I wouldn't have come home. Had I been at the farm, I likely wouldn't have made it to the city hospital in time. As it was, the emergency room was nearly the end of me; my saving grace was that perceptive surgeon who caught what others had missed.

I came back to the land with a body full of fear and pain. Nothing had changed at the farm, but suddenly it was the place where my mind said, "If I'd been out here instead of in the city that day, I wouldn't be alive now." It was months before I could walk properly again or climb into bed without hurt.

Our first drive out to the farm following my surgery was a revelation. It's perhaps clichéd to say I was looking at the world with new eyes at that point, but I really was. There's a certain clarity of vision that comes out of deep pain: you look at the world and see things you never expected to see again because you'd given up, because you'd said at some ultimate point in the process, *I can't go on*. On that first return to the farm after more than a month of recovery, I spent the hour-and-a-half journey northwest from Edmonton watching the scenery change outside the car with this altered view. I found myself very attuned to the way in which the city sprawls its unkempt limbs over more and more of the surrounding farmland; the way, in early April, the light is mellowing and opening into spring, but the land remains clenched like a fist under the snow. The trees know better; despite the gorgeous light, nobody trusts the weather yet.

I felt a similar clenching in my own gut as I shifted uncomfortably in the passenger seat, still unable to sit for long periods of time without great pain, still having to walk with my hands over my abdomen, feeling as though I was about to turn inside out at any moment. I think I realized intellectually at that point, watching the city slowly give way to the last vestiges of parkland and then the black spruce forests of the North, that I could come back to the land deeply fearful, or I could choose to see it as a place of healing. Pain colours us; we carry it behind our eyes for a long time after it's passed. At some point, we have to decide whether we're willing to let it take over our lives and change them permanently, or whether we're going to wrench ourselves open again to the world. I couldn't make that choice in the city. The demands of work and the constant noise kept me stressed and fearful. Thomas and I had always run to the farm when we were facing hurt, so we turned north.

After surgery, I came back to the land with a body that felt nothing like my own. My abdominal muscles had been severed and reattached; for the longest time, I used to laugh to Thomas that I ought to get a tattoo of a zipper pull sketched in at the end of the long, segmented scar up my belly. I couldn't lift anything heavier than a soup can, and the rest of the muscles in my body had begun wasting over the weeks and months of slow recovery.

Before the surgery, I had been a runner. Running was my meditation and my escape, and until I landed in that hospital bed, I was doing fifty to seventy-five kilometres a week on parkland trails. Suddenly, post surgery, I found myself unable even to stand for long periods of time, let alone run. I survived the end of the university term with the concerted, deeply caring help of my family and my students, who carried my books and papers and helped me to walk from one classroom to another until the term was out. And then I found myself at the farm, and it was spring, and that market garden wasn't going to plant itself.

The cabin, that minuscule home, became the centre of my world as I came back to everything I had known. The surgery I had undergone was scarcely ever performed on people my age, the specialist told me, so there was no program of physiotherapy assigned to me after I was released from the hospital. "Give it six weeks," I was told, "and then resume your life." *Resume your life.* Six weeks later, I still couldn't lift a grocery bag without help.

The farm became my therapy while I came back to my body. The greenhouse needed to be set up for the year, the compost needed sieving, the tomatoes needed to be tended and tied in. On cold nights, the cabin was miserable unless we'd chopped wood for the stove. It was the necessity of those tasks and the rightness of their rhythms, done at my body's speed as it gained ground each day, that brought me back. And it was the cabin that held me in the simple vault of its pine walls while I healed. It took nearly two years before the fear left my body. I had been sporadically conscious during the fourteen hours I'd spent on the emergency ward, and during that time, what I had felt was my body giving up. Until the specialist arrived, nobody could diagnose what was wrong, so I was on constant medication for the pain. I was trapped in triage and my heart was beginning to shut down from the drugs in my system. That pain and fear remained huge in me, close under the surface. They were invisible to others, but they had me cold and helpless, even months after the surgery was behind me.

The space that I could have died in became the space that I healed in instead. Dawn to dusk, I watched the light change in the garden. I walked in the forest as I learned to move without pain again. I put my hands in the earth, swung the axe as I was able and slept in the pine-scented loft when my healing body was worn out from the day. It felt right, learning to come back to myself in a place that could easily have held death for me.

When Wallace Stegner speaks, in *Wolf Willlow*, of the southern prairies he knew so well, it is the necessary balance of light and dark that captivates him. "The drama of this landscape is in the sky," he says, "pouring with light and always moving. The earth is passive. And yet the beauty I am struck by, both as present fact and revived memory, is a fusion: this sky would not be so spectacular without this earth to change and glow and darken under it. And whatever the sky may do, however the earth is shaken or darkened, the Euclidean perfection abides."[3] I came to heal in the place that would have killed me, and the cabin held me between its high, bright ribs while I came to terms with my fear.

There is no magic recipe for a farm. There is no script that admits ahead of time who might succeed and who might fail. It's a story that's open to everybody. One thing I can absolutely affirm is that it's a story that will align body and spirit and land. The farm is still the place I go to when I have anything to work through. The cabin is where I will always go to heal, to find my way back.

It's where I'm writing this book.

CHAPTER THIRTEEN

Grow Your Own Home: A Cereal Project

WE INSTANTLY CAME TO LOVE our little cabin – its bright pine and wainscotting; the large, warm sleeping loft; the cast-iron stove on its stone hearth – but we still nursed a fascination for building something with our own hands and at our own pace. Although we sidelined the plan to build a room of our own in order to get a working year-round space on the farm as soon as possible, we never quite abandoned the dream. But more than that, we wanted to build with materials that grew on the land around us. It would be a challenge, as well as a firm link between ourselves and the land we were looking after. The cabin became the perfect stopgap measure while we developed what we soon came to call the "Grow Your Own Home Project."

It's not that we're rebelling against technology out at the farm. Daily, we use the car battery to charge our cellphones and my laptop so that I can write (I rarely do so by hand anymore, except correspondence), and we post pictures of the farm online, where friends and family around the world follow our adventures on this little patch of northern ground. We use gas for the chainsaw and the walking tractor (essentially a Rototiller on steroids). But we have learned that there is great pleasure to be found in doing well with less, especially when it comes to technology.

In the early days of the farm, I became fascinated with Eric Brende's book *Better Off*. A graduate of MIT, Brende moved with his new wife to an Amish community to explore the point at which technology becomes more of a burden than a help in our lives. Because I was teaching in the city during the week and needed the Internet as much for my own research as for availability to my students, I was wondering how much technology was too much, and why exactly these time-saving tools were beginning to feel as though they were absorbing my life. Eric Brende's book drew me from the outset:

> The conviction was growing in me that the besetting prob-
> lem was our culture's blindness to the distinction between
> the tool and the automatic machine. Everyone tended to
> treat them alike, as neutral agents of human interaction. But
> machines clearly were not neutral or inert objects. They
> were complex fuel-consuming entities with certain definite
> proclivities and needs. Besides often depriving their users
> of skills and physical exercise, they created new and artifi-
> cial demands – for fuel, space, money, and time. These in
> turn crowded out other important human pursuits, like
> involvement in family and community, or even the process
> of thinking itself. The very act of accepting the machine was
> becoming automatic.[1]

Although we have a brushcutter to help trim the verges and keep the nettles down along the seismic road out to the hayfield, I've always preferred quieter tools in the garden. If it's possible to do something by hand where before we might have reached for a noisier tool, we gladly put in the additional time. Wheat is harvested by sickle; trees intended for firewood are felled and bucked with the chainsaw, but the logs are chopped by hand with the splitter and the kindling hatchet, and we can listen to the birds while we do it. We asked Mark to design the cabin for a low-voltage electrical system, the same as most recreational vehicles, which can easily be run off a solar panel or two. He put in a handful of electric lights for us, but to this day, three years after the cabin's arrival on the farm, not a single one of them has ever been turned on. We discovered early in the life of the cabin the pleasures of a mellower kind of light, and through the long winter months, our small home is rich with the burnt-sugar scent of beeswax candles. There is no need on this land to reach for large machinery first when the human body can do the same task just as well. The additional time such tasks require allows us to observe the land and to act with deliberation.

Our farm is at the southernmost limits of the boreal, and so we have trees, and lots of them, but they're not the old-growth forests of the Pacific coast. On our quarter, growing on the deep peat bogs, we have muskeg spruce that are fifty years old and yet only ten feet tall and not quite five inches in diameter. It's difficult to find enough consistently sized timber to build a frame house here. But we do have an excellent growing medium: cover cropped yearly to renew its biomass and improve its tilth, the peat yields spectacular crops (though it has the unfortunate trait of continually decomposing, as it's pure organic matter instead of mineral soil). As we looked at the earth, we started thinking, *Why not grow our own home?* That's how we turned to straw bale design.

WE'D BEEN FASCINATED with the wonky straw bale cottages of Europe for many years, but straw bale design hadn't been as visible on the prairies of western Canada. *Why not?* we wondered. Thick bale walls would provide built-in insulation against the wind chill of the Prairie provinces, and the roofline of a bale house could be adjusted to make the most of low winter sunlight for passive heating. But across the Prairies through the early 1900s, inexpensive flat-packed housing could be ordered up from the likes of the Sears catalogue and shipped in via the railway to provide a quick start to a farmstead. This was the route that many farmers chose to follow, and straw bale housing fell by the wayside. As with so many of the projects at our farm, we fell in love with the history and the narrative behind the concept, from the early clay-and-straw cob houses of 1400s Britain to the modern-day straw bale homes going up across North America, and we became new adherents to the concept of self-built, off-grid homes.

One thing we'd learned by growing the market garden year to year was that the current varieties of wheat didn't offer the best options for bale houses. The older varieties, grown with long stalks that would once have been used for animal bedding, provided the best-quality material for the large square bales needed for the house walls. I'd been buying seeds from Jim Ternier at Prairie Garden Seeds in Saskatchewan for years. I was seriously hooked after a trip one spring to St. Peter's Abbey in Muenster, Saskatchewan, where Ternier grows a show garden featuring several varieties of prairie-hardy grains. Not only could we strive to grow and build our own house straight from the land, but we could do it with a minimum of waste. While we used the stalks for the straw bales, we'd choose a selection of wheat known to be decent for baking. The heads of the grain could go to flour while the straw bulk built our house.

We chose Huron wheat, a variety bred in 1925 that originated from a cross made in 1888 in Ottawa. After our first summer on the farm, which had been bone dry, we'd noted that each subsequent year brought months when the garden struggled from constant rain. Huron promised to be a solid mid-season wheat that would stand up to the cooler, damp days of these new summers and deliver a good yield. It was no longer grown in large quantities on the Prairies, so we started up our supply with nothing more than a ten-gram packet of seed from Jim.

That first summer, we brought in ten times what we'd sown. The next summer, we increased the yield exponentially again. Although the stems were long and prone to lodging (falling over) in high winds and rain, Huron was tough. It withstood our late frosts in June and our early frosts in August. It survived hailstorms and grasshoppers. Three years in, we had so much wheat that harvesting the heads and threshing them by hand, as we'd been doing, was no longer an option. We were starting to enter the big time, and at the same point, we needed some practical knowledge to enable us to make the project happen. Cue Trimline Design Centre in Edmonton and a fantastic course about straw bale houses, during which we got the chance to heft and place bales, pegging them into place with bamboo and stitching the wall together. We came out of the course feeling solid about the basics, excited about our project and impatient to start.

Of course, most folk who build straw bale houses buy the bales from a supplier when they're ready to raise a house; they don't single-handedly try to revive a historic wheat variety along the way. Years began to pass, and the ground required for the Grow Your Own Home Project increased by leaps and bounds. We cleared additional plots inside the moose fence that circled our market garden. We brushcut old thickets of wild raspberry canes outside the fence to increase our planting area. Now, several years into the project and with a harvest of over twenty-five pounds of wheat per year, we've got our eyes on the twenty-five-acre hayfield we currently lease to our neighbours for their dairy operation. Huron wheat is tough and grows well; even in a year with early frosts through August and September, the wheat pushes through and ripens off. Although it lodges a bit in the windstorms of summer, it never topples so completely that it fails to ripen. After a few years of growing, I've come to think that there are few lovelier views than a plot of Huron ripening tall and gold and graceful in the late-summer light. The heritage varieties move like dancers in the wind, and where shorter types break under weight, Huron bends fluidly.

So here we are, several years into the project. In another handful of summers, we'll have to take over a section of the hayfield with a tractor to sow out all the Huron we've grown and saved. Twenty-five acres of wheat should give us all the bales we'll need for a working straw bale house. In the meantime, we'll begin the search for timber in the deep woods. The nice thing about load-bearing straw bale construction on a piece of land like ours without too many huge trees is that the building won't need much timber: the bales can hold their own weight up to a storey and a half. So it will be mostly our own Huron that holds up the house, with a scant framing of white spruce where needed. While we build up our seed stock, the straw is handy for stooks to cover woodpiles, and Thomas is dreaming of an archery range with homegrown straw targets.

I love every instance in which we can be self-sufficient, from growing a garden to last us through the cold months to heating the cabin with a wood stove and eschewing propane as frequently as possible. Growing our own house seems like the ultimate project for sustainable farming: a crop and a home in one, and the potential to bring back into common cultivation an intriguing old wheat variety with a lengthy history of its own. It's also wonderful to inhabit the farm's expanding narrative. Although it will take a handful of additional years, like all of the best projects at the farm, the house will make for a fantastic story.

CHAPTER FOURTEEN

The Birds

I was the solitary plover
a pencil
    for a wing-bone
From the secret notes
I must tilt
– Lorine Niedecker, "Paean to Place"

BIRDS ARE THE FIRST THING that sells me on a place. It's because I have more than a little of my paternal grandmother in me, the way she always had a bird feeder close by the window to watch the small comings and goings of the day. There's something indescribable about those tiny lives, simultaneously ephemeral and so deeply present. I instantly root in a place upon learning their names and songs.

Our farm hosts a varying cast of characters as the seasons change. Located at the turning point between parkland and boreal, the forest on our quarter annually resounds with Arctic birds migrating south, gemlike hummingbirds quick-timing it north and all sorts of irruptive species, the good-time guys, here one year and gone the next until the conditions once again meet their champagne tastes.

One of the first elements of the farm we came to know as we worked to create the market garden was the birds. As the ecosystem in the garden area shifted from dead willow thicket to verge landscape with forest edge, the variety of birds broadened. Who knows how they find these places, how they can uncannily locate a landscape in flux and decide that that's going to be a good home in its final iteration, but within a couple of years, our garden transitioned from nothing but chickadees and woodpeckers to an incredible pageant of visitors.

I track our seasons at the farm by the birds. In the winters, it's only the steadfast little holdouts that bear with us through the bitter cold that freezes the garden solid (and the car, too, if we don't go out in shifts to start it overnight). I still don't know how the chickadees make it through those glassine days, but there they are at the feeders when the sun first struggles up behind the muskeg spruce around eleven o'clock in the morning in December.

The ravens stay through the winter too, sweeping in low to check out what's going on around the cabin when we head outdoors during the few bright hours to chainsaw some brush or bring in a load of birch for the stove. They've become so accustomed to us that they no longer greet our car with their *plonk, plonk* that resembles stones dropped into a clear pool, but they're certain to come in low and examine any vehicles they don't recognize. The ravens are uncannily intelligent, devoted to each other and their offspring. Watching them, one has the curious sense of being watched in return; I'm sure they recognize our routines and our every move around the farm and market garden. Their aerobatics on early spring days are our first sign that the cold is breaking.

The pine siskins and waxwings find us in late winter, too, with their trademark calls sizzling like small lasers through the spruces. The siskins are bullies at the feeders, stopping by in pint-sized armies and pushing the chickadees off the black oil sunflower seeds. The waxwings prefer the deep woods, where they spend the small light of the days filching catkins off the birch trees and complaining about the cold. And with the siskins come the grosbeaks, the shy pine variety, each gal carefully escorted by her fella. The showy evening grosbeaks appear in the early spring with their yellow eye markings stamped onto grey feathers like thunderbolts onto some generic superhero. The grosbeaks and siskins are opportunists, though, and they'll stay up north when the birch catkins and spruce cones have a good year. Depending on the availability of food sources, there are gaps in our avian clock as one set of irruptives or another decides to stay where the pickings are best.

On the tails of the grosbeaks and the siskins come redpolls and snow buntings, tiny flocks flashing in the bright air of February on their way up to the Arctic. The sandhill cranes follow them. We know that the spine of winter has well and truly broken when we hear the advance guard of cranes go over, so high as to be almost imperceptible, except that we've been listening for it since the ice went out on the pond. They're heading up to Wood Buffalo National Park and the Peace–Athabasca Delta, their summer breeding grounds, but every spring, like clockwork, the flocks come over and circle the cabin. I've long thought that there are few things lovelier than a large flock of cranes in flight, and years on, the sight still moves me to tears. It's as though you can feel the warmer weather returning in their wake, in the pause that follows their high, warbling calls.

And then there are the Canada geese, making their way north again from the United States, where they've spent the winter in flocks big enough to be considered pests in many areas. I love the geese in spring in a more hopeful way than I do those going out in the autumn. Flying with their mouths open, the spring geese bring a constant, welcome barrage of sound through the April nights. They gather on the Paddle River one quarter over, and on the wildfowl lake up the range road, making the dusks loud with the unutterably joyous havoc of early spring.

The real stunners make their appearance in late April and early May, bringing the warmer summer weather with them, brightening the garden with flashy plumage and fluting calls. We see goldfinches only occasionally, so sunflower yellow that we almost doubt our eyes, and so shy that we're more likely to witness the feeders swinging in their wake and just an afterimage of gold. We're never sure when these lovelies come and go, but it's sometime between April and July, right about when the white-crowned sparrows and hermit thrushes return, their liquid song filigreeing the warming night air.

Last spring, for the first time, we found the May air alive with a new song: reminiscent of woodpeckers, the nasal, keening call repeated itself in sustained bursts throughout the night. We knew we had owls nesting in the forest, mainly great horned owls that came in to pick off the snowshoe hares that pepper the property and the voles growing fat from dropped sunflower seeds at the feeders. This, though, was a sound unlike any we'd heard before. We spent days with bird books and recordings, reading the reports of northern ornithologists online, before we recognized what it was. Northern pygmy owls had come to the farm, and it was with a deep pleasure every night afterward that we listened to their mating calls in the spruce forest. There's nothing quite like the realization that a creature has decided that the land is safe enough and bears food enough to support young. We had four northern pygmy owls this past spring, and we know that come May, we'll be outside again at dusk, listening for their return.

THE NUTHATCHES AND woodpeckers and juncos move into the deep woods over the summer to raise their broods, and while the chickadees are absent, the hummingbirds move in. When we began clearing the dead willow thicket for the market garden, we didn't see a single hummingbird; after planting tall blue spires of delphinium and the surreal red tubes of four o'clocks, as well as growing scarlet runner beans on every trellis, we now see them every year. They build their tiny moss-lined nests in the scrub willow behind the greenhouse, and I can hear them buzzing about as I work to tie up the tomato plants. They've come to know us well enough that we can work a few feet away from the feeders in the garden, but what we love about the hummingbirds, like all the birds that visit the farm, is that they are always essentially wild. Not a single chickadee gets seeds from a hand. Feeders, yes, but I have no desire to get in their space. The fact that they choose to spend part of the summer on the farm is reward enough.

The blue jays are perennial on the farm, and during the occasional summer, they'll be joined by a pair of whiskey jacks who raise a brood of silent little grey shadows in the market garden. Unlike their more rambunctious cousins, the whiskey jacks are relatively tongue-tied, but the young ones love to play just as much as their brighter counterparts. We put saucers of water out for them to splash in, and we groan over the strawberries, each of which, if we're not careful, receives a tiny, inquisitive drill hole from a questing little beak.

The red-tailed hawks return in the spring, and there's inevitably a day when we realize that the sharp screams we've been hearing in the garden are no longer the baby jays playing hawk, but the hawks themselves, home and hunting. The mother raises a brood every year in the old poplars fringing the pond, and we see them late in the afternoon once they've fledged, the mother on point and the babies cavorting after, snapping at dragonflies.

We'd only just noticed a new song in the garden two summers ago and recognized the tail-bobbing antics of a marsh wren when we heard a rattling in the stove box one afternoon. It was a cool summer day, the first we hadn't needed to light the stove, though we were contemplating it. The rattling in the box threw me off: what on earth had defeated the fire baffles and gone down the flue? When I popped open the door, there was a tiny wren, ash-blind and terrified. She was just small enough to have threaded her way down the chimney, and had the fire been lit, that would have proven the end of her. Thomas caught her up in a gloved hand and carefully released her from the front porch.

We'd put up some nesting boxes in the flower garden that spring, and it was about a week later that we realized one of them had a tenant. It was the stove wren, rattling away in a different sort of box, bringing in small twigs and grasses and bits of thistledown for her nest. Each year since her rescue from the stove, she's raised a brood or two in the garden. I wrote a poem in memory of her near escape, and each spring, when we see the nesting box rustle on its post and hear her pert little song, I read it over again. It's a private ceremony of sorts as the avian clock shifts to warmer weather and the winter relaxes its grip on the North Country at last.

The Stove Wren
(after Claudia Emerson's "Waxwing")

A skitter of claws in the chimney flue,
  she plummets into the stove box,
    angles a breadth of wing against the glass.

On any other night but this,
  she'd have made her plunge
    to a fire laid and lit,

sap-thick tamarack, patient birch,
  or chafed her death unnoticed
    in the cold cinders of a deserted room.

Her unexpected entrance
  thwarts the baffles, iron and firebrick.
    Ash-blind, she trusts my hands,

tympani of small claws, heart
  a racing pocketwatch against
    the slim feathers of her breast.

She seizes the rain-washed air
  with all her wings, leaves my hands
    empty, fingertips burning

with ash,
  with absence. The true
    measure of her leaving

that small space
  once cupped and thrumming
    between my palms.

CHAPTER FIFTEEN

Mishaps and Miracles

WE CAN ALMOST BALANCE OUT, one for one, the disasters and glories on the farm. Things happen that scare the pants off us, and then we turn around and witness the evening light in July painting itself across the ripening wheat field out front, and we catch ourselves thinking that nothing could possibly be finer.

We're pretty careful on the farm around the machinery. There are no unnecessary risks (believe you me, when we finally lay our hands on a baler for the Grow Your Own Home Project, nobody is going to be hauling out a jam from that thing while it's running). But machinery is fallible, and every once in a while, something very nearly goes spectacularly wrong. Something like running a compound mitre saw off an emergency generator and suddenly realizing the saw pulls so much juice that it pops the fuse on the generator every time you fire it up. Or watching a valve break on the fuel line of the walking tractor, turning a formerly essential piece of farm machinery into a diesel version of a Super Soaker. But sometimes it's human error.

Tiredness is a major enemy of industry (and safety) on the farm. Let's face it: it's downright enervating to work full-time off the farm during the week, teaching the kids we love so much and marking their work all hours of the night, and then to drive up to the farm each weekend and work again there. It's a choice, absolutely, and we make it gladly to keep our farm alive and flourishing, and also because we can't imagine our lives without teaching, but sometimes it has a cost. I remember a time when Thomas came in from a Saturday afternoon of mid-winter tree-felling after a hard week with his fourth-graders, his face white as a sheet and a good-sized tear out of his orange chainsaw chaps. When I started squawking, fingering the tangled protective threading sticking out of the tear, he told me, "I knew it was time to call it a day when I finished sawing and dropped the saw to rest it on my thigh. Totally forgot that the blade was still winding down!" If it hadn't been for his chainsaw chaps, filled with long fibres designed to grab the saw teeth and slow them down, his career as a farmer (not to mention his stature) might have been pretty short indeed.

There are times, too, when what goes wrong has nothing to do with anything human. Downed trees are a personal favourite of mine, those fence-bouncers that come rocketing down and seem to be trying their darnedest to take out the moose fence around the market garden in one fell swoop. Sometimes they're occasioned by a bad drop with the chainsaw or a sudden leading wind (in spite of a felling wedge), but just as often, the trees have heart rot and come down on their own, usually within reach of something smashable. There's nothing worse than lying in the cabin at night during a windstorm and hearing the telltale crack of a tree splitting its top.

We keep any snag trees near the cabin felled for safety's sake, but our area is prone to the occasional spectacular windstorm, and those have a tree-moving power all their own.

We were working in the yard one summer afternoon when a windstorm started blowing its way up. Engrossed in gate-making, we hadn't noticed the sky turning a bruised black and the sun vanishing behind a veil of distant rain. The first crack of thunder almost overhead caught us by surprise.

"Let's pack up and head into the tool shed to get this gate done," Thomas suggested, and it was just as well we went. The tool shed, a former wooden flour bin from a bakery in town, had been dragged out to the property by the old farmer who'd sold us the land. Painted an eye-popping blue that I'm sure threatens the sight of pilots of low-flying aircraft, it's the place we go to work on overcast days or afternoons of fierce sun. We'd been in the tool shed for all of ten minutes when the thunder really started cracking and the lightning started coming down like some sort of firework display.

We don't have many large trees on the property, just a few big old spruces down by the hayfield and one massive poplar by the front gate. From our place, the boreal gets shorter and shorter as it shifts from mixed woods to small, dark spruce running north to the tundra. But that lightning went like nobody's business for the old sentry poplar out by the front gate. The blast corkscrewed down the tree into the ground around it and blew the bark right off; it hit the tool shed with a thump like cannon fire.

It was akin to being inside a drum when it's sounded – the shed resounded with the strike. We didn't dare step outside until we were certain, five or ten minutes later, that the centre of the storm had passed over and the ground was no longer electrified. When we did, it was with ringing heads and shot hearing. The poplar was still standing, scarred and green, but in the following years it would die, along with several of the trees around it. We kept it standing as long as we could – the scarred trunk was a good story-starter – until we started to get a little worried about it dropping its dead weight onto the fence or the camper. Down it came, but the story lives large in our memories. We still scoot for cover the moment a lightning storm looms.

Quite aside from the mishaps, though, there are the times when everything comes together in astonishing beauty. There are days in the spring when the sandhill cranes fly over to their breeding grounds at the far northern tip of the province, the size of the flocks growing annually, warbling away at the craziest altitudes, almost out of sight. There's something ethereal about the cranes; the way they catch the thermals, high and delicate as rice-paper kites, never ceases to astonish. And there are the small beauties: mowing out the seismic road to the hayfield and having to stop to move nests of tiny snowshoe hares off into the spruce forest. Folk can tell me I'm waxing lyrical as much as they want, but something about cupping those minuscule lives in my hands, warm from the sun and damp from their small fear, sparks a groundedness in me that I can't find in the city.

We've come to know it as "Farm TV" over the years. The market garden calls for an awful lot of work, usually when we're already exhausted from a week at another job, but we've learned to make time to sit and watch the light change or the cranes fly. We've begun to venerate the way a cow moose will move out of the trees in the spring, pared to the bone by winter, pocked with tick marks and weary in every slant of her body, and the bright-eyed slink of the new calf following at her heels through the early willow.

The tiniest things give you that feeling on the land, as though your chest is seizing and you have to stop and just watch whatever happens until the bit of beauty that's caught you decides to let you go.

One of the most astonishing sights I've witnessed on the farm has to be the dragonfly hatch each summer. In early July, when the pond has warmed up just enough for the mosquitoes to be flying dawn to dusk, the dragonflies appear magically, almost overnight. Although the deep wood still possesses its own resonant supersonic hum, the air in the market garden is now alive with crisp wings. Two, three hundred dragonflies take to the air early in the day and don't settle again until evening, when it's safer to be hidden amongst the leaves than out where the little brown bats can find them. In late July, the air crackles with something akin to electricity as hundreds of dragonfly wings stroke a cappella; in early August, the young hawks fledge and practise hunting in the air above the garden. We sit entranced, watching the rolling aerobatics as the small birds clip dragonflies from the sky with precision. Later, weeding amongst the cabbages and lettuce, we'll find disarticulated dragonfly wings scattered and catching the light.

For every moment that stops the heart with panic, there's a balancing moment of beauty. I won't wax poetic about the land in a perfectionist sense: we work hard out here, and things constantly threaten the tiny equilibrium we've established in the market garden. Whatever peace we find is often hard won. But I stand firmly with Berry and Kingsolver and so many other writers who possess a deep need to step outside the city to find a place of calm. I don't like the word "authentic"; at best, it's divisive and antagonistic, implying one way of being is intrinsically better than another. But I do very much favour the notion of *alignment*. I'm convinced that at the heart of the matter lies a desire to draw what we do into alignment with how we live. Some of us aren't in a place where we can live consistently on the land that holds our hearts, but come mishaps or miracles, we're bound and determined to make that land as much a part of who we are as humanly possible.

CHAPTER SIXTEEN

The Soapbox, Please: Mediated Spaces and Nature as a Speaking Subject

Systems are like wet rawhide; when they dry, they strangle what they bind.
– Wallace Stegner, letter to Cheryll Glotfelty[1]

I have no interest in a "return to Nature," which strikes me as an especially decadent form of aestheticism, like an adult of forty pretending to have the innocence of a child.
– Harold Fromm, "From Transcendence to Obsolescence: A Route Map"[2]

I'VE COME TO UNDERSTAND that the natural world, as far as modern politics go, will always be rife with contradictions. I think this is partly because for so many thousands of years, the natural world was made mute; unvoiced, insentient and thus subservient. Humans, with our logic and our ability to think and reason, ended up higher in the heap through a self-imposed, religion-based hierarchy that preferred brainpower above any other adaptation for survival. The implications for the rest of the world, as we made this tremendous leap in our own estimation, were enormous: "Nature *is* silent in [Western] culture (and in literate societies generally) in the sense that the status of being a speaking subject is jealously guarded as an exclusively human prerogative."[3] Nature, as it became unvoiced and unspeaking, also became a tool, something lacking its own internal drive.

As a young woman who names herself both teacher and farmer, I have been called out time and again by those who want to stamp me as a back-to-the-lander, hipster, hobby farmer, weekender. I came to believe early on in the entire process that terms like these (and the list is long) are not only derogatory in the assumptions they make about my devotion to the land and commitment to my farm, but an oversimplification when applied to the natural world itself. There's something a little shameful lurking behind the term "back-to-the-lander," as though nature is nothing more than sunsets and meteor showers and forests, struck dumb and demanding articulation through human perception. As though, by going back to the land, there's something childlike about me and others of the same ilk.

As a farmer, I am sometimes driven off the land by the madness of an increasingly changeable set of weather patterns, but that is not the only reason I go. I find the other half of my meaning, my justification, in the classroom. It's here, in our humanities classrooms, amidst the books and the spoken narratives of our many cultures, that I believe the major issues of environmental politics have one of the greatest potentials for exploration. We deal in ideas, after all. We gather and process not only data but stories. As Gwendolyn MacEwen explained in her poem "Dark Pines Under Water," "there's something down there and you want it told."[4] There is a great deal "down there" in our collective Western consciousness when it comes to the environment and the baggage of the word "nature," especially when personified as "Nature," and much of that has to do with our reluctance to give up our perceived power.

"To regard nature as alive and articulate has consequences in the realm of social practices. It conditions what passes for knowledge about nature and how institutions put that knowledge to use,"[5] Christopher Manes argues. We are working our way toward a vision of a rights-wielding (and speaking) nature that we are comfortable with, and although we're not at the finish line yet, the race has been started. We may not always be sure how to handle an empowered natural world, but we're beginning to address some of the reasons for our distancing from nature. Novelists such as Ruth Ozeki are taking on GMO crops and the ethics of the meat industry, while Mark Frankland explores the human narrative and the environmental costs of widespread livestock diseases such as foot-and-mouth and BSE. Canadian writer and critic Di Brandt's *So this is the world & here I am in it* explores the thin boundaries between spirit, land and human technology imposed by the Mennonite culture. These are only a few of the many fiction writers and poets tackling issues of nature, agriculture and health; the list is not exhaustive, and every year these books become more of an international presence. People are creating powerful art and literature that questions our distance from the natural world around us.

The unvoicing of the natural world is too often interpreted as a clash of culture and nature, which, as far as I can see it, is a contrast akin to a Saskatchewan back road in early spring: prone to gumbo and utter bog-down. Humans don't know which way to jump when it comes to choosing between nature and culture, partly because we think we *have* to choose. As Frederick Turner writes:

More often than need be, Americans [and I would add
Canadians to the mix here] confronted with a natural landscape have either exploited it or designated it a wilderness
area. The polluter and the ecology freak are two faces of the
same coin; they both perpetuate a theory about nature that
allows no alternative to raping it or tying it up in a plastic
bag to protect it from contamination.[6]

Of course, increasingly in Canada, and most visibly in Alberta's Oil
Patch, we are oversubscribing to the theory of "make hay while the
sun shines" and damn the human and environmental consequences.

So where is one to fall on the continuum of human versus
environment? At its basis, this is a discussion of semiotics, the sign-
making (and thus meaning-making) of the natural world and the
human world. I teach and research ecocriticism, though my own
standing is well within the camp of ecosemiotics, blending both cultural semiotics (a human-based dialogue where the natural world
enters the discussion only as a touchstone, a referent) and biological semiotics (where signification occurs in the wild world whether
or not we as humans are conscious of it). Think of the two in this
way: cultural semiotics is akin to looking at a perfectly manicured
garden, pruned and sprayed within an inch of its life, and reading
the human intent first, with the natural world coming a distant second. Nature is only there in that garden as a touchstone, a tip of
the hat to the natural world in a place that is very structured and
controlled. Biosemiotics is like looking at a bird that has evolved a
certain behaviour to confuse predators and lure them away from a
nest on the ground. The predator sees the flopping bird and pursues
it, drawn away from the nest in the process. This is meaning-making
that occurs quite outside the snooping of humans; it is happening

every day around us in the natural world, in the interactions between plants and animals.

I don't believe that all environmental phenomena are, at their essence, possessed of intent; instead, the more I interact with the land at our farm, the more I subscribe to the middle ground. Not everything in nature happens for a reason, and natural occurrences aren't meaningfully tied to the will of some higher being(s), but I do see the natural world making signs, making decisions – responding, evolving, adapting those responses. In this way, the natural world is not so far removed from the human at all – and thus I think we have *more*, not less, responsibility in our treatment of both wild and mediated spaces.

I lean toward the thinking of Charles Sanders Peirce, for whom nature and culture are not diametrically opposed; rather, both show evidence of mind and so have a common root. In times when the environment is under duress, this commonality can act as a bridge between two areas often seen as radically different.[7] It's no longer acceptable to come at the natural world from a wholly humanistic or wholly environmental view; the interlinked systems are more complicated than that. And notice I said "*interlinked*," because I fundamentally believe that the human and natural worlds are points on the same continuum, not blips on two tracks running parallel to one another, destined only occasionally to conflict with static-like bursts of friction leaping the lines.

Ecosemiotics, at its best and most complete, brings together the cultural and the biological, saying that the natural world is capable of mind or intent, and that we humans interact with environments capable of independent meaning-making, of which we are only one interpreting participant. Ecosemiotics has as its goal, as Riste Keskpaik says, "to help to diminish communication problems between human and nature, because from that [ecocritical] viewpoint, it becomes possible to speak about nature, as it seems to us in culture, and to speak with nature, because its ability of speech has been restored."[8]

You might wonder how a field of study turned into a way of life and a rather unconventional off-grid small farm. Let's confront it head on: too often, our environmental politics focus on nature as either something that can be pillaged or something that is mindless, in a beautiful or threatening way. It becomes something entirely *other* from the human world, and so it's easier for us to dodge it or wreck it with little harm to our consciences. In the case of Alberta's Oil Patch, we're aware that great harm is being done, but that's what natural resources are *for*, right? Development and profit. And hey, we rehabilitate the land later and turn it back into parkland, so it's all okay in the end, right?

Wrong. It's not all cultural semiotics; the entire meaning of the natural world can't be filtered through human experience and human history. But it's not all biosemiotics, either: nature isn't simply lurking out there, running its own scheme, intangible and unknowable. We live in a world of mediated spaces, where even the words "nature" and "wild" have a variety of connotations. Is the land around Fort McMurray *wild*? Is a park wild? What about an urban forest – is that still nature? We have to get comfortable in our language, culture, actions and economy with the idea of nature as both wild and mediated space. We need to work with the concept of ecosemiotics, that nature's meaning is composed partly of the natural world and partly of how we interpret it. It's not one or the other; we have to meet somewhere in the middle.

I want to talk about these mediated, ecosemiotic spaces in two regards: nature writing and gardens. If you've made it this far in the book, you've probably spotted that my intent is both to acknowledge the beauty, power and meaning-making of the natural world and to explore my place as an organic small farmer in a much larger non-organic framework. "By changing individual experiences of the author to become a part of wider experience of culture," Timo Maran says, "nature writing becomes a strategy for regarding and valuing nature. Writing about nature is simultaneously a recognition that nature as such is worth writing and talking about."[9] I like what Maran has to say, as he gives some examples that I can relate to, that I see in place daily out on my farm:

> In many cases the living activities of different organisms merge in the environment in a way that makes it very difficult to distinguish the contributions of different species in it. As such, nature becomes a medium or *interface*, which different living beings read and where they write into. The example of such collective creation of environment is forest. Life cycles of different organisms in forests combine in complicated ways; some species form habitats for others, the decay of some organisms becomes food and source material for others, and so on. Forest is full of information and communicative relationships, which [...] brings up the question of how people read forests, what aspects of it they are able to interpret and how.[10]

I would extend this idea, as Maran does in his other essays, to include the human–nature interface that is a garden. I look again at our farm garden, which is a mediated space between the natural and the manmade worlds. It adapts what we see as most successful "beyond the fence" in the kinds of crops we plant, how we build our beds and where we place certain crops to best access sun or dodge frost. I see our garden surrounded each summer by wild carrot, or Queen Anne's lace, so I know that a domesticated crop from the same Apiaceae family, our regular orange carrot, will do well in our

soil. The flowers I plant are carefully chosen to draw in natural pol-
linators or domestic bees: domesticated asters in the garden mimic
the wild asters that grow in profusion outside the fence. Pulling from
human culture, the Memory Gardens draw on the concept of flowers
being evocative of certain people. Both the human and the natural
are essential components of our farm garden.

The mediated space – a cottage garden, an urban forest, a
sustainably harvested forest – is, like the nature text, often underesti-
mated. If it's done well, we don't look at the mediated space and see
a clunky conglomeration of wild and domesticated flowers splat-
tered together. We become aware of a number of systems working
together, simultaneously, to the benefit of each system.

The politics of the environment – of our air, our water, our earth
and our food – are so often inflammatory. How can they not be?
We're all aware that we're escalating the environmental crisis, and
many groups believe they have the solution: whether it's to stick
one's head in the sand, turn tail and run in survivalist mode or ditch
Big Oil and go instantly green. My own politics align with those of
Maria Rodale: I call for change, as we cannot go on the way we have
been, whether in the West or around the world. Rodale argues that
it's not too late to heal the damage we've done to the planet, but we
need to make changes now, beginning with our agriculture. If we're
going to change the face of agriculture from large-scale business to
small, diversified farms, we ought to go fully organic to restore the
health of our soil and our people. Her approach is very much a sym-
biotic one, where we cultivate the health of the earth while profiting
from it. This seems to me a more viable way forward than our current uncertainty of how to interact with our environment: we either
rubber-stamp it as inviolate or wreck it to the bone. Our fear and
misunderstanding show clearly in the contradictory behaviours we
are capable of.

MY CALL IS to the mediated space. We can't change our dependence on Big Oil and Big Ag overnight. It's not reasonable to ask people to give up their modern way of life, any more than it is reasonable to ask them to give up all of their technology. It invites fear and rejection. But what we *can* do is invite them to participate in mediated spaces, in the creation of interfaces between ourselves and the wild world. Instead of wholesale clear-cutting, selective logging. Instead of Big Ag, small farms and greater awareness of ecosystems. When we put in the effort to make those changes, we need to make them count and turn to organic practices as much as we can. It makes no sense to intervene on behalf of the natural world with a giant vat of chemicals.

This sort of widespread change won't happen overnight. But we in the West recovered from what we did to the land during the Dust Bowl, when millions of acres of good soil blew away because of wholesale land clearing. We can recover from what we've done to Earth this time, too. Maria Rodale explains that we haven't yet lost this fight, and we have the chance to face the future differently:

> We still have time to heal the planet, feed the world, and keep us all safe. But we have to start right now. To implement this change, we must unite to fight the most important fight of our lives, perhaps one of the most important global struggles in the history of our species. It's a fight for survival. Because no matter what our political beliefs, our religion, our family values, our sexual preferences, our tastes in music or foods are, we are all in this together. Our fates are linked.[11]

To her call for action, I'll add that "we" has to mean "humans plus natural world." We need to change the way in which we interact with the land around us in a fundamental way. Our spaces under threat need to become mediated spaces. Our every interaction with an increasingly threatened environment needs to become one where we meet at an interface. Overt destruction can no longer be the norm, because we very truly are in this together – every living thing. We can't take back the damage we've inflicted, but with effort, I do believe we can change it, as long as we are willing to give up our notion of being at the top of the power pyramid. Let's face it: things are dire, and there is no hierarchy in this struggle for survival.

CHAPTER SEVENTEEN

A Fine Line between Loving and Leaving

Go out in the woods, go out. If you don't go out in the woods, nothing will ever happen and your life will never begin.
– Clarissa Pinkola Estés, *Women Who Run With the Wolves*

"HOW CAN YOU CLAIM TO love your farm so much," I'm often asked, "when you live away from it for such long periods of time?"

Let me tell you something about feeling a compass behind your breastbone every minute of every day, its needle constantly tugging you north and west, away from the prairie proper and into the hem of the boreal.

Let me tell you about being attuned to the sound of geese passing over in all weather, the way the air changes when it rains. You can sense it, even inside the layers of brick and glass, the building that separates you from the land dissolving as the scent of wild, wet earth creeps down the corridor.

It doesn't take a farm to invoke the iron taste of leaving in your mouth. Anyone who loves a small plot of ground – a city garden, a vacant lot with some guerilla beds, a balcony of pots – understands the almost physical hurt of parting from it, even for a minor stint. I hurt every day I wake up in our city bed, wondering how the light will be changing over the front field or across the pond, whether the moose will be in the willow by the cabin again, if the wren has fledged her young ones yet and we'll return to find the box untenanted. I can feel where the farm is at any point in my day, not out of some arcane sixth sense developed from years of summer nights out there with the coyotes under the stars, but because of the bond between that earth and this body. Some grounds we choose; some are our instinctive homes.

I leave because I love this land too much. It's a seeming contradiction, I know, and one I face over and over again at each leave-taking. Part of the reason is financial: we both work off the farm so that we can afford to keep our land debt-free. The larger reason, though, is that we want to offer something back. We find ourselves in the incredible position of caring for a piece of land that is unique in a multitude of ways, nearly untouched, unspoiled in a province where much of the wild land goes to oil and gas. We work off the farm because we want to show our students the necessity of watching out for the land; as teachers, we both feel that that's as big a part of our jobs as the daily curriculum. Learn well, be kind to each other, care for where you live.

When my third book of poetry, *Seldom Seen Road*, came out and began to be reviewed, I had to get used to the notion of being considered an idealist. I was writing about the land in a way that didn't always focus, in every poem, every line, on the environmental harm being done to my province. Yes, I despair for what is happening to the land around me, but I find beauty in the land's ability to survive, too. I struggle every day with the knowledge of what Big Oil is doing, but I also see the strength of the land around me and think it vital to celebrate. I sat uncomfortably with the label of idealist for a while before I realized that the twisting in my gut came not from my disagreeing with the term, but from my questioning why that should be a wholly negative thing.

I believe many of us who long to have our paws in the dirt are, in one manner or another, idealists. I think we all secretly share the notion that the more people who can care for a piece of land, whether that's a community garden plot or a tomato in a bucket out on the balcony, the more likely it is that those same people will think carefully about what happens to the land around them. It's next to impossible to work with the earth and not take a hit when that land is treated with disrespect.

Maybe I'm an idealist because I think that it's not too late, that students actually want to understand the impacts of their choices on their own landscapes. Everybody has a story of home, and at its core, home is a plot of land, no matter how nebulous our life stories might be after that initial belonging. After years of fearing that I wouldn't be able to sustain our small family on my sessional instructor's salary after Thomas retired, I took a tenure-track job at Red Deer College, three hundred kilometres south of our farm, to teach creative writing and ecocriticism to the children of farmers and Oil Patch workers. My new and beloved job takes me on a weekly trek far from the piece of land I call home because I really believe in talking about our sense of place, and that stories are one of our first shared lines of defense of the places we care about.

"Go out in the woods, go out," Estés says. We've done that, Thomas and I, quite literally. There is no calmer place for me than walking the forest trails on the land we call home. And we've gone into the other kind of woods, too: the city, where we are much less at ease, but where we go in hopes of sharing with young people some stories about this piece of ground and helping them connect to their own intrinsic sense of place, of belonging.

Perhaps the farm itself is my biggest nod to idealism, to lightening the pessimistic baggage I carry when considering the fate of wild spaces in my province. We have no children of our own; that's not the way our lives will go, although we spend most of our days caring for the children of others. The farm is our most significant idealistic leap: the thought that one day, we'll turn this land over to young people as devoted to it as we have been, and the hope that they too will possess the desire beyond all others to keep safe this small piece of ground.

I CANNOT THINK of the boreal without experiencing an almost physical ache for the hundreds and hundreds of acres destroyed by open-pit mines, leaky gas pipelines and oil wells. But I also can't think of that land without feeling that old pull behind my breastbone: the very real feeling of homing, of return. How two people from away can come to love a mad little patch of the northern Canadian forest so deeply continues to be a surprising mystery to me, but ultimately, I think it's more about respect than birthplace. Respect eases many boundaries. Fundamentally, it's about the privilege and responsibility of caring for a piece of land in a time when every bit counts.

As I've been writing this, the harvest has been coming in. Through the fading light of the weekends, we've been pulling the last of the potatoes and beets, planting out the fall garlic under an insulating blanket of leaves and tucking the final rows of carrots under fleece blankets to sweeten against the frost and early snow. During the weekdays, I watch from my office window at the college and think long and hard about how best to draw out the care I find in so many young people for the land. I heft the weight of that job at a time when much of the funding in my province comes from oil and gas, industries that offer open doors to young people at the expense of the landscapes those plants and processors move into. The price for all of us is so high.

An idealist? If that means carrying the twin burdens of fear and love, then yes, I'll accept the term. I'm terrified for the future of the land in my province for the children of the students I now teach. I want them to be able to hear the cranes go over with the piping of bone flutes in spring. I want them to know what it means to love a piece of land so much that you'll leave it, over and over, to do the necessary work of creating respect for that same earth.

I'm wondering how to approach the coming year's new eco-criticism class at the college. Maybe I ought to start out with some basic critical theory about environment, landscape and place – John Muir and Thoreau, a bit of Rachel Carson. Perhaps we should watch a movie to get the students immediately immersed in the sorts of landscape we'll be discussing. Or what about a brief foray into some readings by various Canadian environmental critics and ecologists?

There's this place I know, I find myself writing, *three hundred kilometres northwest of where we're standing right now. No matter where I am, I can always tell you where that land is; it's as though there's a line as the crow flies that connects me to it every time I have to be away.* I can imagine pausing at this point and looking up around a classroom full of kids from across the province, some from smaller prairie cities, some from towns that no longer exist on the map in any way more substantial than by name. *Let's talk about place,* I imagine saying. *Let's talk about the landscapes we belong to. For me, it all starts with this mad little farm on the edge of the boreal forest. No matter where I go, I will always return to that ground.*

I want to know about the land you belong to. And so we'll do two introductions on this first day, as we learn who we are and where we stand. I want to know about you, and I want to know about the land you carry with you.

Think about it for a moment. Quietly, deeply. Feel your feet on that ground, the quality of the air and light. When you're centred, anchored in body and in memory within that space, I want you to take us there.

Ready?

Go.

Who Shall I Say is Calling?: A Eulogy to Place

Once there were brook trout in the streams in the mountains. You could see them standing in the amber current where the white edges of their fins wimpled softly in the flow. They smelled of moss in your hand. Polished and muscular and torsional. On their backs were vermiculate patterns that were maps of the world in its becoming. Maps and mazes. Of a thing which could not be put back. Not be made right again. In the deep glens where they lived all things were older than man and they hummed of mystery.
– Cormac McCarthy, *The Road*

And who by fire, who by water,
. . . .
And who shall I say is calling?
– Leonard Cohen, "Who by Fire"

I. Who by Earth

Say Fort Chipewyan. Say Mikisew Cree and Athabasca Chipewyan Nations. Say Lake Athabasca, Mackenzie River delta. Say Wabasca, say Peace Country, say Chinchaga–Rainbow. Say Burnaby Mountain.

Go to the heart of it. Say Manning and Spirit River, Swan Hills, High Level, Grande Cache. Confront a tainted blood of pipelines and pumpjacks, proposed nuclear-reactor sites, open-pit mines. Say Beaverlodge and Fox Creek, Fairview, Nampa, Donnelly, Hythe. Hydrogen sulfide flames mapping a different set of constellations.

What we teach when we cannot keep the drill sites off our land.

What we teach when our air is rank with flaring and our water lights from fracking gases.

What we teach is helplessness.

II. Who by Water

Name our lifelines, bloodlines. Blood thicker than water; water slick with effluvium, oil sheen. Name the Petitcodiac. Name the East-main and the Rupert, the Okanagan, the Taku, the Iskut. Name the Groundhog, the Milk, the Bow. Name the Peel. Name the Red River. Name the Churchill.

Say watershed. Say transboundary. Say Joint Commission.

Name what moves in our veins. Say biochemical oxygen demand. Say eutrophication. Say phosphorus and nitrogen, say dead zone. Say petroleum and polycyclic aromatic hydrocarbons. Say lead. Mercury. Sulphur dioxide. Phthalates.

Name the Yamaska, trolling Montérégie farmland. Name the Bayonne, the Don. Track the livestock runoff bruised-blue algae blooms until they can be seen from space.

Name the Elk and the Fording, selenium skeins through cutthroat trout, frog eggs, red-winged blackbirds.

No Salmon of Knowledge, Fionn mac Cumhaill: in the Cariboo, sing Hazeltine, Quesnel, Fraser. Sing out the breach at Mount Polley and the salmon sloughing their skin in tailings-pond runoff.

III. Who by Fire

Say Windsor–Detroit. Say particle pollution. Say smog alert.

Say asthma. Say eight million and rising. Oh, Canada.

Say sulphur dioxide, nitrogen oxide, neurotoxic contamination.

Say shale gas and Elsipogtog. Say Liard and Cordova basins. Say compressor stations and seismic cutlines. Say Horn River Basin.

Say exploration wells, say finger-dipping in the Klappan, the Kootenays. Say the mines and the coalfields: Fording River, Greenhills,

Line Creek. Bullmoose. Quinsam. Say Tuya River. Say Coal River. Say Bowron River and Hat Creek.

Say horizontal drilling, say abandoned wellbores. Say leaking and flaring, say surface-water allocation. Say open-ended.

Say red-eyed vireo. Magnolia warbler. Thrush. Rose-breasted grosbeak. Say seventy-five hundred immolated in a St. John's gas flare.

Say   conservative estimate.

NOTES

Preamble

1   Paul Hawken, "Commencement Address to the Class of 2009, University of Portland" in *Hope Beneath Our Feet: Restoring Our Place in the Natural World* (Berkeley: North Atlantic Books, 2010), 4.

Chapter One

1   Michael Pollan, *A Place of My Own* (New York: Penguin, 2008), 4.

2   Barbara Kingsolver, "Called Home" in *Animal, Vegetable, Miracle: A Year of Food Life* (Toronto: Harper Collins, 2008), 3.

Chapter Three

1   Wendell Berry, *The Art of the Commonplace* (Berkley: Counterpoint Press, 2002), xviii.

2   Ibid., xiv.

3   Wendell Berry, quoted in Norman Wirzba, "The Challenge of Berry's Agrarian Vision" in *The Art of the Commonplace: The Agrarian Essays of Wendell Berry*, ed. Norman Wirzba (Berkeley: Counterpoint, 2002), x.

Chapter Four

1   Wendell Berry, "A Native Hill," in *The Art of the Commonplace* (Berkeley: Counterpoint Press, 2002), 7.

2   Ibid., 8.

Chapter Five

1   Lisa Martin-DeMoor, "One last time, in our old kitchen," in *One Crow Sorrow* (Edmonton: Brindle & Glass, 2008), 7.

Chapter Eight

1   Barbara Langhorst, *Restless White Fields* (Edmonton: NeWest Press, 2012), 44.

Chapter Twelve

1   Wendell Berry, *The Art of the Commonplace* (Berkeley: Counterpoint, 2002), 118.
2   Michael Pollan, *A Place of My Own* (New York: Penguin, 2008), 4.
3   Wallace Stegner, *Wolf Willow* (New York: Penguin, 1990), 7.

Chapter Thirteen

1   Eric Brende, *Better Off* (New York: HarperCollins, 2005), 6.

Chapter Sixteen

1   Quoted in Cheryll Glotfelty, "Introduction: Literary Studies in an Age of Environmental Crisis," in *The Ecocriticism Reader: Landmarks in Literary Ecology*, edited by Cheryll Glotfelty and Harold Fromm (Athens: University of Georgia Press, 1996), xxii.
2   Harold Fromm, "From Transcendence to Obsolescence: A Route Map" in *The Ecocriticism Reader: Landmarks in Literary Ecology*, edited by Cheryll Glotfelty and Harold Fromm (Athens: University of Georgia Press, 1996), 33.
3   Christopher Manes, "Nature and Silence" in *The Ecocriticism Reader: Landmarks in Literary Ecology*, eds. Cheryll Glotfelty and Harold Fromm (Athens: University of Georgia Press, 1996), 15-16.

4 Gwendolyn MacEwen, "Dark Pines Under Water" in *The Shadow Maker* (Toronto: Macmillan, 1972), 50.

5 Christopher Manes, "Nature and Silence" in *The Ecocriticism Reader: Landmarks in Literary Ecology,* eds. Cheryll Glotfelty and Harold Fromm (Athens: University of Georgia Press, 1996), 15–16.

6 Frederick Turner, "Cultivating the American Garden" in *The Ecocriticism Reader: Landmarks in Literary Ecology,* eds. Cheryll Glotfelty and Harold Fromm (Athens: University of Georgia Press, 1996), 45.

7 Charles Sanders Peirce, from Winfried Nöth, "Ecosemiotics and the Semiotics of Nature," *Sign Systems Studies* 29, no.1 (2001): 75.

8 Riste Keskpaik, quoted in Timo Maran, "Towards an Integrated Metholodology of Ecosemiotics: The Concept of Nature-Text," *Sign Systems Studies* 35, no.1/2 (2007): 279.

9 Timo Maran, "Towards an Integrated Metholodology of Ecosemiotics: The Concept of Nature-Text," *Sign Systems Studies* 35, no.1/2 (2007): 288.

10 Ibid., 285.

11 Maria Rodale, *Organic Manifesto: How Organic Food Can Heal Our Planet, Feed the World, and Keep Us Safe* (New York: Rodale, Inc., 2010), xvi.

ACKNOWLEDGEMENTS

GRATEFUL THANKS TO THOSE who have supported and inspired the development of this manuscript, most particularly Noelle Allen at Wolsak and Wynn for her enthusiasm and deep interest in the project; the marvellous team of Emily Dockrill Jones, Ashley Hisson, Paul Vermeersch and Joe Stacey for their talent, patience and time in making the book come to life; and Marijke Friesen for the beautiful cover design. Thanks to Jacqueline Baker at MacEwan University for the invitation to speak as part of the Canadian Authors Series in 2013; Ariel Gordon and the Under Western Skies Conference at Mount Royal University, where some of the essays from this book received their first public reading in 2014; the University of Alberta Devonian Botanic Garden; Trimline Design Centre in Edmonton; the Alberta Foundation for the Arts; and LitFest Alberta (Fawnda Mithrush).

Thanks also to the wonderful people who have, in a myriad of unique ways, made our farm the place it is today through their time, expertise and generosity: Mark Carpenter; Bruno Wiskel; Les and Laura Tywoniuk; Patty Milligan; Thean Pheh; Stan Galloway; Michael Gravel (MG Creative); Richard and Yasmin Butler; and the many friends and family members who have brought their support and their joy to this ground.

BIBLIOGRAPHY

Bennett, Jennifer. *The New Northern Gardener.* Willowdale, ON: Firefly Books, 1996.

Berry, Wendell. *The Art of the Commonplace.* Berkeley: Counterpoint Press, 2002.

——. *The Unsettling of America: Culture & Agriculture*, 3rd ed. San Francisco: Sierra Club Books, 1996.

Brandt, Di. *So this is the world & here I am in it.* Edmonton: NeWest Press, 2007.

Brende, Eric. *Better Off.* New York: HarperCollins, 2005.

Brett, Brian. *Trauma Farm: A Rebel History of Rural Life.* Vancouver: Greystone Books, 2009.

Bubel, Mike, and Nancy Bubel. *Root Cellaring: Natural Cold Storage of Fruits and Vegetables.* North Adams, MA: Storey Publishing, 1991.

Buell, Lawrence. *The Environmental Imagination: Thoreau, Nature Writing, and the Formation of American Culture.* Cambridge, MA: Harvard University Press, 1995.

Butala, Sharon. *The Perfection of the Morning: An Apprenticeship in Nature.* Toronto: HarperCollins, 2005.

——. *Wild Stone Heart: An Apprentice in the Fields.* Toronto: Harper-Collins, 2000.

Chiras, Daniel D. *The Solar House: Passive Heating and Cooling.* White River Junction, VT: Chelsea Green Publishing, 2002.

Cohen, Leonard. "Who by Fire." www.azlyrics.com/lyrics/leonard-cohen/whobyfire.html.

Cooley, Dennis. *Perishable Light.* Regina: Coteau Books, 1988.

Crump, Jeff, and Bettina Schormann. *Earth to Table: Seasonal Recipes from an Organic Farm.* Toronto: Random House, 2009.

Coleman, Eliot. *Four-Season Harvest.* White River Junction, VT: Chelsea Green Publishing, 1999.

——. *The New Organic Grower.* White River Junction, VT: Chelsea Green Publishing, 1995.

Davies, Gareth, and Margi Lennartsson, eds. *Organic Vegetable Production: A Complete Guide.* Ramsbury, UK: The Crowood Press, 2005.

Elpel, Thomas J. *Living Homes: Integrated Design and Construction.* Pony, MT: HOPS Press, 1990.

Estés, Clarissa Pinkola. *Women Who Run With the Wolves: Myths and Stories of the Wild Woman Archetype.* Toronto: Random House, 1995.

Fossel, Peter V. *Organic Farming: Everything You Need to Know.* Minneapolis: Voyageur Press, 2007.

Gidley, Mick, and Robert Lawson-Peebles, eds. *Views of American Landscapes.* Cambridge, UK: Cambridge University Press, 1989.

Glotfelty, Cheryll, and Harold Fromm, eds. *The Ecocriticism Reader: Landmarks in Literary Ecology.* Athens: University of Georgia Press, 1996.

Goodbody, Axel, and Kate Rigby, eds. *Ecocritical Theory: New European Approaches.* Charlottesville: University of Virginia Press, 2011.

Gopnik, Adam. *Winter: Five Windows on the Season.* Toronto: House of Anansi, 2011.

Hellum, A. K. *Listening to Trees.* Edmonton: NeWest Press, 2008.

Herriott, Carolyn. *A Year on the Garden Path: A 52-Week Organic Gardening Guide.* Victoria, BC: EarthFuture Publications, 2005.

——. *The Zero-Mile Diet: A Year-Round Guide to Growing Organic Food.* Madeira Park, BC: Harbour Publishing, 2010.

Kennedy, Des. *The Passionate Gardener: Adventures of an Ardent Green Thumb*. Vancouver: Greystone Books, 2006.

Keogh, Martin, ed. *Hope Beneath Our Feet: Restoring Our Place in the Natural World*. Berkeley: North Atlantic Books, 2010.

Kingsolver, Barbara. *Animal, Vegetable, Miracle*. Toronto: Harper-Collins, 2008.

Kull, Kalevi. "Semiosphere and a Dual Ecology: Paradoxes of Communication." *Sign Systems Studies* 33, no. 1 (2005): 175–89.

Lacinski, Paul, and Michel Bergeron. *Serious Straw Bale: A Home Construction Guide for All Climates*. White River Junction, VT: Chelsea Green Publishing, 2000.

Langhorst, Barbara. *Restless White Fields*. Edmonton: NeWest Press, 2012.

Lawrence, James M. *The Harrowsmith Country Life Reader*. Charlotte, NC: Camden House Publishing, 1990.

Lima, Patrick. *The Harrowsmith Perennial Garden*. Camden East, ON: Camden House Publishing, 1989.

——. *The Organic Home Garden: How to Grow Fruits and Vegetables Naturally*. Buffalo: Firefly Books, 2004.

Lindström, Kati, Kalevi Kull, and Hannes Palang. "Semiotic Study of Landscapes: An Overview from Semiology to Ecosemiotics." *Sign Systems Studies* 39, no. 2/4 (2011): 12–36.

MacEwen, Gwendolyn. *The Shadow-Maker*. Toronto: Macmillan, 1972.

Maran, Timo. "Gardens and Gardening: An Ecosemiotic View." *Semiotica* 150, no. 1/4 (2004): 119–33.

——. "Towards an Integrated Metholodology of Ecosemiotics: The Concept of Nature-Text." *Sign Systems Studies* 35, no. 1/2 (2007): 269–94.

Martin-DeMoor, Lisa. *One Crow Sorrow*. Edmonton: Brindle & Glass, 2008.

McCarthy, Cormac. *The Road.* London: Vintage, 2007.

Murakami, Haruki. *After Dark.* London: Vintage, 2003.

Nearing, Helen, and Scott Nearing. *The Good Life: Helen and Scott Nearing's Sixty Years of Self-Sufficient Living.* New York: Schocken, 1990.

Nöth, Winfried. "Ecosemiotics and the Semiotics of Nature." *Sign Systems Studies* 29, no. 1 (2001): 71–81.

Pheh, Thean. *Fruits, Nuts, and Berries for Edible Landscaping in Alberta.* Edmonton: Trees N More, 2013.

Pitzer, Sara. *Homegrown Whole Grains.* North Adams, MA: Storey Publishing, 2009.

Pollan, Michael. *The Botany of Desire: A Plant's-Eye View of the World.* Toronto: Random House, 2002.

——. *A Place of My Own.* New York: Penguin, 2008.

——. *Second Nature: A Gardener's Education.* New York: Grove Press, 1991.

Rodale, Maria. *Organic Manifesto: How Organic Food Can Heal Our Planet, Feed the World, and Keep Us Safe.* New York: Rodale, 2010.

Savage, Candace. *A Geography of Blood: Unearthing Memory from a Prairie Landscape.* Vancouver: Greystone Books, 2012.

Schwenke, Karl. *Successful Small-Scale Farming: An Organic Approach.* North Adams, MA: Storey Publishing, 1991.

Seymour, John. *The Self-Sufficient Life and How to Live It.* New York: DK Publishing, 2003.

Smith, Alisa, and J.B. MacKinnon. *The 100-Mile Diet: A Year of Local Eating.* Toronto: Vintage Canada, 2007.

Solomon, Steve. *Gardening When It Counts: Growing Food in Hard Times.* Gabriola Island, BC: New Society Publishers, 2005.

Steen, Bill, Athena Swentzell Steen, and Wayne J. Bingham. *Small Strawbale: Natural Homes, Projects, and Designs.* Salt Lake City: Gibbs Smith, 2005.

Stegner, Wallace. *Wolf Willow.* New York: Penguin, 1990.

Thoreau, Henry David. *Walden.* New York: Penguin, 1986.

Whitman, Walt. *Leaves of Grass.* Toronto: Ryerson Press, 1965.

Willis, Elizabeth, ed. *Radical Vernacular: Lorine Niedecker and the Poetics of Place.* Iowa City: University of Iowa Press, 2008.

Wirzba, Norman. "The Challenge of Berry's Agrarian Vision." In *The Art of the Commonplace: The Agrarian Essays of Wendell Berry,* edited by Norman Wirzba. Berkeley: Counterpoint, 2002.

Wiskel, Bruno. *Designing and Landscaping the Family Home.* Colinton, AB: Evergreen Environmental, 2002.

——. *Woodlot Management.* Edmonton: Lone Pine Publishing, 1995.

Woginrich, Jenna. *Made From Scratch: Discovering the Pleasures of a Handmade Life.* North Adams, MA: Storey Publishing, 2008.

C. W. HILL

JENNA BUTLER is the author of three books of poetry and ten short collections with small presses. Butler teaches creative writing and eco-criticism at Red Deer College. In the summer, she and her husband live on a small organic farm near the historic Grizzly Trail in Alberta's north country.